WHAT HAVE
YOU CHANGED
YOUR MIND ABOUT?

BOOKS BY JOHN BROCKMAN

AS AUTHOR
By the Late John Brockman
37
Afterwords
The Third Culture: Beyond the Scientific Revolution
Digerati

AS EDITOR
About Bateson
Speculations
Doing Science
Ways of Knowing
Creativity
The Greatest Inventions of the Past 2,000 Years
The Next Fifty Years
The New Humanists
Curious Minds
What We Believe But Cannot Prove
My Einstein
Intelligent Thought
What Is Your Dangerous Idea?
What Are You Optimistic About?
Science at the Edge

AS COEDITOR
How Things Are

WHAT HAVE YOU CHANGED YOUR MIND ABOUT?

TODAY'S LEADING MINDS RETHINK EVERYTHING

EDITED BY JOHN BROCKMAN

WITH AN INTRODUCTION BY BRIAN ENO

HARPER ● PERENNIAL

NEW YORK ● LONDON ● TORONTO ● SYDNEY ● NEW DELHI ● AUCKLAND

FIRST EDITION

Designed by Aline C. Pace

Library of Congress Cataloging-in-Publication Data is available upon request.

ISBN 978-0-06-168654-2

09 10 11 12 13   OV/RRD   10 9 8 7 6 5 4 3 2 1

CONTENTS

⎯⎯◇⎯⎯

PREFACE

◄◦►

The *Edge* Question

In 1991, I suggested the idea of a third culture, which "consists of those scientists and other thinkers in the empirical world who, through their work and expository writing, are taking the place of the traditional intellectual in rendering visible the deeper meanings of our lives, redefining who and what we are." By 1997, the growth of the Internet had allowed implementation of a home for the third culture on the Web, on a site named *Edge* (www. edge.org).

Edge is a celebration of the ideas of the third culture, an exhibition of this new community of intellectuals in action. They present their work, their ideas, and comment about the work and ideas of third culture thinkers. They do so with the understanding that they are to be challenged. What emerges is rigorous discussion concerning crucial issues of the digital age in a highly charged atmosphere where "thinking smart" prevails over the anesthesiology of wisdom.

The ideas presented on *Edge* are speculative; they represent the frontiers in such areas as evolutionary biology, genetics, computer science, neurophysiology, psychology, and physics. Some of the fundamental questions posed are: Where did the universe come from? Where did life come from? Where did the mind come from? Emerging out of the third culture is a new natural philosophy, new ways of understanding physical systems, new ways of thinking that call into question many of our basic assumptions of who we are, of what it means to be human.

An annual feature of *Edge* is The World Question Center, which was introduced in 1971 as a conceptual art project by my friend and collaborator, the late artist James Lee Byars. His plan was to gather the hundred most brilliant minds in the world together in a room, lock them in, and "have them ask each other the questions they were asking themselves." The result was to be a synthesis of all thought. Between idea and execution, however, are many pitfalls. Byars identified his hundred most brilliant minds, called each of them, and asked them what questions they were asking themselves. The result: Seventy people hung up on him.

But by 1997, the Internet and e-mail had allowed for a serious implementation of Byars's grand design, and this resulted in our launching *Edge*. For each of the anniversary editions of *Edge*, I have used the interrogative myself and asked contributors for their responses to a question that comes to me, or to one of my correspondents, in the middle of the night.

The 2008 *Edge* Question:
When thinking changes your mind, that's philosophy. When God changes your mind, that's faith. When facts change your mind, that's science.

What have you changed your mind about? Why?

Science is based on evidence. What happens when the data change? How have scientific findings or arguments changed your mind?

John Brockman
Publisher & Editor, *Edge*

INTRODUCTION

———◦❯———

Brian Eno

There is now an almost total disconnection between the validity of a story and its media success. If it's a good enough—or convenient enough—story, it will echo eternally around the media universe. We lack any publicly accepted way of saying "This is demonstrably wrong," and as a result there is almost no disincentive to unconstrained spinning, trafficking in poor information, and downright lying. The result is a diminishing accountability at almost every level of public discourse and a burgeoning industry of professional Swiftboaters.

There used to be a regular program on BBC radio during the 1980s. It was only five minutes long, but in that five minutes the makers sought to examine a modern myth to see whether it held up to scrutiny. During those Reagan/Thatcher years, a popular way of attacking the conspicuously successful (and inconveniently socialistic) Swedish social system was to knowingly point out that they had the world's highest suicide rate—as though the price of all that official altruism was widespread cultural despondency. Actually it turned out that Sweden didn't have an especially high suicide rate—the country was ranked somewhere in the thirties, below France, Spain, Japan, Belgium, Austria, Switzerland, Denmark, and Germany, and just one place above the United States—but it was too good a story to drop, and to this day you'll still hear it.

I suppose it doesn't really matter if people continue to think the Swedes are killing themselves at record-breaking rates, or if people believe that *The Da Vinci Code* is a true story (as I was confidently assured by a New York policeman), or if they think that Eskimos have four hundred words for snow, but it really does matter if they believe that Saddam Hussein was responsible for 9/11, or that global warming is an anticapitalist plot, or that Senator Kerry was a coward during his Vietnam service. These things matter because they have direct real-world consequences, and in a media-soaked universe they point out the Achilles' heel of democracy. Democracy was intended to flourish by engaging the intelligence of the wider population—on the assumption that people might, on average, be able to assess what is in their best interests. But if the information upon which they make their assessments is of poor quality, how can that work?

This question is more urgent as we blunder from one global crisis to another on the basis of bad or spun information. We need to be confident of the data we're using and to know when we lack trustworthy data. And we need to act upon reliable data rather than sidelining it if it doesn't sit comfortably with the agenda to which we've already committed. We need, in short, a way of arriving at positions based more on knowledge and reason than on ideology, political convenience, or the needs of business. And we need, in particular, to move on from unproductive positions—to know and admit when we're wrong.

We have one great example of a sort of cultural conversation where this is the case, and that is science. Whatever its shortcomings and perversions (and many of these essays draw attention to them), science is nonetheless committed to an explicit system of validation. Scientific statements are made in accessible (though not always easy) language and propose hypotheses that must be testable—that is, subject to a comparison with current evidence and logic. Among other things, this means that people will sometimes have to admit that an idea they have cherished and nurtured is, in fact, wrong. This act of pragmatic humility is the very underpinning of science and one of the prime reasons for its successes. It allows us to move from ideas that are clearly untrue toward ones that are, at least for now, serviceable. At a certain point, we all had to agree that the world could usefully be regarded as round.

Science is an extraordinary intellectual invention, a construction designed to neutralize the universal human tendency to see what we expected to see and overlook what we didn't expect. It is designed in many ways to *subtract* data—to create experiments sufficiently insulated from externalities that we can trace a clear connection from a cause to an effect. It seeks to cancel out prejudices, environmental contingencies, groupthink, tradition, pride, hierarchy, dogma—to subtract them all and to see what this or that corner of the world looks like without their effects.

But then there is the brain itself—a brilliant machine but with a few occasionally troublesome work-arounds. These work-arounds enable us to form complex ideas from insubstantial data—to "jump to conclusions" or "make intuitive leaps." They allow us to recognize familiar patterns when we see only parts of them. They allow us to make deductions about meaning from context. All these valuable talents carry with them a downside: A familiar but wrong conclusion is more likely to be reached than an unfamiliar one.

Most of us, in our daily lives, arrive at our feelings about things through a vast and usually unexamined hotchpotch of received opinion, prejudice, cultural consensus, personal experience, reason, and observation. We would generally be hard-pressed to trace the source of those feelings in any rigorous way: Why, for instance, would I call myself a liberal pragmatist rather than, say, a neoconservative or a Trotskyite? I'd like to think it's because I carefully evaluated the various philosophical choices. But I know that I'm not free of prejudice (those neocons *dress* so badly!) and that I probably arrived at my current position as much through my particular unexamined hotchpotch as through reason. If we don't know the sources of the attitudes we hold, we can avoid taking responsibility for them and leave ourselves plenty of ways of saying we weren't really wrong—or that the thing that has turned out to be wrong wasn't what we meant anyway.

Most of the contributors to this volume are scientists or writers about science and are thus people who accept a special kind of responsibility for their ideas and present them as the results of accessible rational processes—in such a way that the ideas can be both understood and reexamined. It sounds straightforward, but many of the writers here are at pains to point out that the process is not as transparent as this description would suggest. We aren't without investments in our ideas—and for academics, ideas are rarely just academic.

The conceit of science is that we should be satisfied when our ideas are shown to be wrong, for we've moved on from a delusion and our picture of the world is therefore less wrong. Of course, it doesn't feel like this at the time; far from knowing more, we feel like we suddenly know a lot less. A bit of the world we thought we'd secured is wild and mysterious again. This is a difficult feeling to live with; it's much more comforting to cling to a familiar idea even when it runs against the tide of evidence than to have no idea at all.

If you've spent a long time thinking yourself into a certain intellectual position, you are naturally resistant to letting it go: A lot of work went into it. If it "felt right" to you for whatever complex mesh of personal reasons makes an idea "feel right," then to abandon it isn't just a question of rationality but also a question of self-esteem. For if *that* feeling was wrong, how many others might be? How much of the rest of your intellectual world are you going to have to pick apart? And if everyone has watched you thinking your way there and seen you building your city around it, there might also be the simple issue of losing face.

For this reason, these essays exhibit a kind of intellectual honesty often lacking in much of human discourse. They give me real hope for the human race. The humility it takes to admit that you're wrong and the fortitude it takes to endure the chilly interval of uncertainty until the next step—finding a better idea—are manifest again and again. And there's courage, too, when evidence leads you away from the comfortable consensus into thoughts that might be highly incendiary and politically sensitive—thoughts you might not even like very much yourself.

If we are ever going to achieve a rational approach to organizing our affairs, we have to dignify the process of admitting to being wrong. It doesn't help matters at all if the media, or your friends, accuse you of "flip-flopping" when you change your mind. Changing our minds is our hope for the future.

CHRIS ANDERSON

Editor in chief of Wired *magazine; Author of* The Long Tail

Seeing Through a Carbon Lens

Aside from whether Apple matters (whoops!), the biggest thing I've changed my mind about is climate change. There was no one thing that convinced me to flip from "wait and see" to "the time for debate is over." Instead, there were three things, which combined for me in early 2006. There was, of course, the scientific evidence, which kept getting more persuasive. There was also economics, and the recognition that moving to alternative, sustainable energy was going to be cheaper over the long run as oil got more expensive. And finally there was geopolitics, with ample evidence of how top-down oil riches destabilized a region and then the world. No one reason was enough to win me over to total energy regime change, but together they seemed win-win-win.

Now I see the entire energy and environmental picture through a carbon lens. It's very clarifying. Put CO_2 above everything else, and suddenly you can make reasonable economic calculations about risks and benefits, without getting caught up in the knotty politics of full-spectrum environmentalism. I was a climate skeptic and now I'm a carbon zealot. I seem to annoy traditional environmentalists just as much, but I like to think that I've moved from behind to in front.

1

BRIAN GOODWIN

Biologist, Schumacher College, U.K.; author of Nature's Due: Healing Our Fragmented Culture

Pan-Sentience

I have changed my mind about the general validity of the mechanical worldview that underlies the modern scientific understanding of natural processes. Trained in biology and mathematics, I have used the scientific approach to explain natural phenomena during most of my career. The basic assumption is that whatever properties and behaviors have emerged naturally during cosmic evolution can all be understood in terms of the motions and interactions of inanimate entities—elementary particles, atoms, molecules, membranes and organelles, cells, organs, organisms, and so on.

Modeling natural processes on the basis of these assumptions has provided explanations for myriad natural phenomena, ranging from planetary motion and electromagnetic phenomena to the properties and behavior of nerve cells and the dynamic patterns that emerge in ant colonies or flocks of birds. There appeared to be no limit to the power of this explanatory procedure, which enchanted me and kept me busy throughout most of my career in biology.

However, I have now come to the conclusion that this method of explaining natural phenomena has serious limitations, and that these come from the basic assumptions on which it is based. The crunch came for me with the "explanation" of qualitative experience in humans and other organisms. By this I mean the experience of pain or pleasure or well-being, or any other of the qualities that are very familiar to us.

These are described as "subjective"—that is, experienced by a living organism—because they cannot be isolated from the subject experiencing them and measured quantitatively. What is often suggested as an explanation of this is evolutionary complexity: When an organism has a nervous system of sufficient complexity, subjective experience and feelings can arise. This implies that something totally new and qualitatively different can emerge from the interaction of "dead," unfeeling components such as cell membranes, molecules, and electrical currents.

But this implies getting something from nothing, which violates what I have learned about emergent properties: There is always a precursor property for any phenomenon, and you cannot just introduce a new dimension into the phase space of your model to explain the result. Qualities are different from quantities and cannot be reduced to them.

So what is the precursor of the subjective experience that evolves in organisms? There must be some property of neurons or membranes or charged ions producing the electrical activity associated with the experience of feeling that emerges in the organism.

One possibility is to acknowledge that the world isn't what modern science assumes it to be—mechanical and "dead"—but that everything has some basic properties relating to experience or feeling. Philosophers and scientists have been down this route before and call this view pan-sentience or panpsychism: the idea that the world is impregnated with some form of feeling in every one of its constituents. This makes it possible for complex organized beings, such as organisms, to develop feelings and for qualities to be as real as quantities.

Pan-sentience shifts science into radically new territory. Science can now be about qualities as well as quantities, helping us to recover quality of life, to heal our relationship to the natural world, and to undo the damage we are causing to the Earth's ability to continue its evolution along with us. It could help us to recover our place as participants in a world that is ours not to control but to contribute to creatively, along with all the other diverse members of our living, feeling planetary society.

SAM HARRIS

Neuroscientist; author of The End of Faith *and* Letter to a Christian Nation

Optimizing Our Design

Like many people, I once trusted in the wisdom of nature. I imagined there were real boundaries between the natural and the artificial, between one species and another, and thought that with the advent of genetic engineering we would be tinkering with life at our peril. I now believe that this romantic view of nature is a stultifying and dangerous mythology.

Every hundred million years or so, an asteroid or comet the size of a mountain smashes into the Earth, killing nearly everything that lives. If ever we needed proof of nature's indifference to the welfare of complex organisms such as ourselves, there it is. The history of life on this planet has been one of merciless destruction and blind, lurching renewal.

The fossil record suggests that individual species survive, on average, between one million and ten million years. The concept of a "species" is misleading, however, and it tempts us to think that we, as *Homo sapiens*, have arrived at some well-defined position in the natural order. The term "species" merely designates a population of organisms that can interbreed and produce fertile offspring; it cannot be aptly applied to the boundaries between species (to what are often called "intermediate" or "transitional" forms). There was, for instance, no first member of the human species, and there are no canonical members now. Life is a continuous flux. Our nonhuman ancestors bred, generation after generation, and incrementally begat what we now deem to be the species *Homo sapiens*—ourselves. There is nothing about our ancestral line or our current biology that dictates how we will evolve in the future. Nothing in the natural order demands that our descendants resemble us in any particular way. Very likely, they will not resemble us. We will almost certainly transform ourselves, likely beyond recognition, in the generations to come.

Will this be a good thing? The question presupposes that we have a viable alternative. But what is the alternative to our taking charge of our biological destiny? Might we be better off just leaving things to the wisdom of nature? I once believed this. But we know that nature has no concern for individuals or for species. Those that survive do so despite her indifference. While the process of natural selection has sculpted our genome to its present state, it has not acted to maximize human happiness, nor has it necessarily conferred any advantage upon us beyond the ability to raise the next generation to child-bearing age. In fact, there may be nothing about human life after the age of forty (the average lifespan until the twentieth century) that has been selected by evolution at all. And with a few exceptions (e.g., the gene for lactose tolerance), we probably haven't adapted to our environment much since the Pleistocene.

But our environment and our needs—to say nothing of our desires—have changed radically in the meantime. We are in many respects ill-suited to the task of building a global civilization. This is not a surprise. From the point of view of evolution, much of human culture, along with its cognitive and emotional underpinnings, must be epiphenomenal. Nature cannot "see" most of what we are doing, or hope to do, and has done nothing to prepare us for many of the challenges we now face.

These concerns cannot be waved aside with adages like "If it ain't broke, don't fix it." There are innumerable perspectives from which our current state of functioning can be aptly described as broke. Speaking personally, it seems to me that everything I do picks out some point on a spectrum of disability: I was always decent at math, for instance, but this is simply to say that I am like a great mathematician who has been gored in the head by a bull; if Tiger Woods awoke from surgery to find that he now possessed (or was possessed by) my golf swing, rest assured that a crushing lawsuit for medical malpractice would be in the offing.

Considering humanity as a whole, there is nothing about natural selection that suggests our optimal design. We are probably not even optimized for the Paleolithic, much less for life in the twenty-first century. And yet we are now acquiring the tools that will enable us to attempt our own optimization. Many people think this project is fraught with risk. But is it riskier than doing nothing? There may be current threats to civilization that we cannot even perceive, much less resolve, at our current level of intelligence. Could any rational strategy be more dangerous than following the whims of nature? This is not to say that our growing capacity to meddle with the human genome couldn't present some moments of Faustian overreach. But our fears on this front must be tempered by a sober understanding of how we got here. Mother Nature is not now, nor has she ever been, looking out for us.

Philosopher, Harvard University; author of Betraying Spinoza

The Popperian Sound Bite

Edge's question this year wittily refers to a way of demarcating science from philosophy and religion: "When thinking changes your mind, that's philosophy. When facts change your mind, that's science." Behind the witticism lies the important claim that science—or more precisely, scientific theories—can be clearly distinguished from all other theories, that scientific theories bear a special mark and what this mark is is falsifiability. Said Karl Popper, the criterion of the scientific status of a theory is its falsifiability.

For most scientists, this is all they need to know about the philosophy of science. It was bracing to come upon such a clear and precise criterion for identifying scientific theories. And it was gratifying to see how Popper used it to discredit the claims that psychoanalysis and Marxism were scientific theories. It long seemed to me that the falsifiability test was basically right and enormously useful.

But then I started to read Popper's work carefully, to teach him in my philosophy of science classes, and to look to scientific

8

practice to see whether his theory survives the test of falsifiability (at least as a description of how successful science gets done). And I've changed my mind.

For one thing, Popper's characterization of how science is practiced—as a cycle of conjecture and refutation—bears little relation to what goes on in the labs and journals. He describes science as if it were skeet-shooting, as if the only goal of science is to prove that one theory after another is false. But just open a copy of *Science*. To pick a random example: "In a probabilistic learning task, A1-allele carriers with reduced dopamine D2 receptor densities learned to avoid actions with negative consequences less efficiently." Not "We tried to falsify the hypothesis that A1 carriers are less efficient learners, and failed." Scientists rarely write the way Popper says they should, and a good Popperian should recognize that the Master may have oversimplified the logic of theory testing.

Also, scientists don't, and shouldn't, jettison a theory as soon as a disconfirming datum comes in. As Francis Crick once said, "Any theory that can account for all of the facts is wrong, because some of the facts are always wrong." Scientists rightly question a datum that appears to falsify an elegant and well-supported theory, and they rightly add assumptions and qualifications and complications to a theory as they learn more about the world. As Imre Lakatos, a less-cited (but more subtle) philosopher of science, points out, all scientific theories are unfalsifiable. The ones we take seriously are those that lead to "progressive" research programs, where a small change accommodates a large swath of past and future data. And the ones we abandon are those that lead to "degenerate" ones, where the theory gets patched and repatched at the same rate as new facts come in.

Another problem with the falsifiability criterion is that it has become a blunt instrument, unthinkingly applied. Popper tried to use it to discredit not only Marxism and Freudianism as sci-

entific theories but also Darwin's theory of natural selection—
a position that only a creationist could hold today. I have seen
scientists claim that major theories in contemporary cosmology
and physics are not "science" because they can't think of a simple test that would falsify them. You'd think that when they are
faced with a conflict between what scientists really do and their
memorized Popperian sound bite about how science ought to be
done, they might question the sound bite and go back and learn
more than a single sentence from the philosophy of science. But
such is the godlike authority of Popper that his is the one theory
that can never be falsified.

Finally, I've come to think that identifying scientificality with
falsifiability lets certain nonscientific theories off the hook, by
saying that we should try to find good reasons to believe whether
a theory is true or false only when that theory is called "science."
It allows believers to protect their pet theories by saying that they
can't be, and shouldn't be, subject to falsification, just because
they're clearly not scientific theories. Take the theory that there's
an omnipotent, omniscient, beneficent God. It may not be a scientific hypothesis, but it seems to me to be eminently falsifiable;
in fact, it seems to have been amply falsified. But because falsifiability is seen as demarcating the scientific, and since theism
is so clearly not scientific, believers in religious ideologies get
a free pass. The same is true for many political ideologies. The
parity between scientific and nonscientific ideas is concealed by
thinking that there's a simple test that distinguishes science from
nonscience, and that that test is falsifiability.

Psychologist and computer scientist; former director of the Institute for the Learning Sciences; author of Dynamic Memory Revisited

Specialized Intelligences

When reporters interviewed me in the 1970s and '80s about the possibilities for Artificial Intelligence I would always say that we would have machines as smart as we are within my lifetime. It seemed a safe answer, since no one could ever tell me I was wrong. But I no longer believe that will happen. One reason is that I am a lot older and we are barely closer to creating smart machines.

I have not soured on AI. I still believe we can create very intelligent machines. But I no longer believe that those machines will be like us. Perhaps it was the movies that led us to believe we would have intelligent robots as companions. (I was certainly influenced by Kubrick's *2001*.) Certainly most AI researchers believed that creating machines that were our intellectual equals or better was a real possibility. Early AI workers sought out intelligent behaviors to focus on—like chess or problem solving— and tried to build machines that could equal human beings in those endeavors. While this was an understandable approach, it was, in retrospect, wrongheaded. Chess playing is not really a typical intelligent human activity. Only some of us are good at it, and it seems to entail a level of cognitive processing that, while impressive, seems quite at odds with what makes humans smart. Chess players are methodical planners. Human beings are not.

Humans are constantly learning. We spend years learning some seemingly simple stuff. Every new experience changes what we know and how we see the world. Getting reminded of our previous experiences helps us process new experiences better than we did the time before. Doing that depends on an unconscious indexing method that people learn to do without quite realizing they're learning it. We spend twenty years (or more) learning how to speak properly and learning how to make good decisions and establish good relationships. But we tend to not know what we know. We can speak properly without knowing how we do it. We don't know how we comprehend. We just do.

All this poses a problem for AI. How can we imitate what humans are doing when humans don't know what they're doing when they do it? This conundrum led to a major failure in AI—expert systems, which relied on rules that were supposed to characterize expert knowledge. But, the major characteristic of experts is that they get faster when they know more, whereas more rules made systems slower. The idea that rules were not at the center of intelligent systems meant that the flaw was relying on specific consciously stated knowledge instead of trying to figure out what people meant when they said they just knew it when they saw it, or they had a gut feeling.

People give reasons for their behaviors, but they are typically figuring that stuff out after the fact. We reason nonconsciously and explain rationally later. Humans dream. There obviously is some important utility in dreaming. Even if we don't understand precisely what the consequences of dreaming are, it's safe to assume that it's an important part of the unconscious reasoning process that drives our decision making. So an intelligent machine would have to dream, because it needed to, and it would have to have intuitions that proved to be good insights, and it would have to have a set of goals that made it see the world in a way that an entity with different goals would not. In other words, it would need a personality, and not one that was artificially installed but one that came with the territory of what it was about as an intelligent entity.

What AI can and should build are intelligent special-purpose entities. (We can call them Specialized Intelligences or SIs.) Smart computers will indeed be created. But they will arrive in the form of SIs, which will make lousy companions but will know every shipping accident that ever happened and why (the shipping industry's SI), or be an expert on sales (a business world SI). The sales SI, because sales is all it ever thinks about, will be able to recite every interesting sales story that ever happened and the lessons to be learned from it. For some salesman about to call on a customer, for example, this SI would be fascinating. We can expect a foreign policy SI that helps future presidents learn about the past in a timely fashion and helps them make decisions—it will know all the decisions the government has ever made and have cleverly indexed them so as to be able to apply what it knows to current situations.

So AI, in the traditional sense, will not happen in my lifetime, nor in my grandson's lifetime. Perhaps a new kind of machine intelligence will one day evolve and be smarter than us, but we are a really long way from that.

Philosopher, University Professor, codirector of the Center for Cognitive Studies, Tufts University; author of Breaking the Spell: Religion as a Natural Phenomenon

What Could a Neuron "Want"?

I've changed my mind about how to handle the homunculus temptation: the almost irresistible urge to install a "little man in the brain" to be the Boss, the Central Meaner, the Enjoyer of pleasures, and the Sufferer of pains. In *Brainstorms* (1978), I described and defended the classic GOFAI (Good Old-Fashioned AI) strategy that came to be known as "homuncular functionalism," replacing the little man with a committee:

> *The AI programmer begins with an intentionally characterized problem, and thus frankly views the computer anthropomorphically: if he solves the problem he will say he has designed a computer than can [e.g.,] understand questions in English. His first and highest level of design breaks the computer down into subsystems, each of which is given intentionally characterized tasks; he composes a flow chart of evaluators, rememberers, discriminators, overseers and the like. These are homunculi with a ven-*

geance. . . . Each homunculus in turn is analyzed into smaller homunculi, but, more important, into less clever homunculi. When the level is reached where the homunculi are no more than adders and subtractors, by the time they need only the intelligence to pick the larger of two numbers when directed to, they have been reduced to functionaries "who can be replaced by a machine." (p. 80)

I still think this is basically right, but I have recently come to regret—and reject—some of the connotations of two of the terms I used: "committee" and "machine." The cooperative bureaucracy suggested by the former, with its clear reporting relationships (an image enhanced by the no-nonsense flow charts of classical cognitive science models), was fine for the sorts of computer hardware—and also the levels of software, the virtual machines—that embodied GOFAI, but it suggested a sort of efficiency that was profoundly unbiological. And while I am still happy to insist that an individual neuron, like those adders and subtractors in the silicon computer, "can be replaced by a machine," neurons are bio-machines profoundly unlike computer components in several regards.

Notice that computers have been designed to keep needs and job performance almost entirely independent. Down in the hardware, the electric power is doled out evenhandedly and abundantly; no circuit risks starving. At the software level, a benevolent scheduler doles out machine cycles to whatever process has highest priority, and although there may be a bidding mechanism of one sort or another that determines which processes get priority, this is an orderly queue, not a struggle for life. (As Marx would have it, "From each according to his abilities, to each according to his needs.") It is a dim appreciation of this fact that probably underlies the common folk intuition that a computer could never "care" about anything. Not because it is made out of the wrong materials—why should silicon be any less suitable a substrate for caring than organic molecules?—but because its internal economy has no built-in risks or opportunities, so it doesn't have to care.

Neurons, I have come to believe, are not like this. My mistake was that I had stopped the finite regress of homunculi at least one step too early! The general run of the cells that compose our bodies are probably just willing slaves—rather like the selfless, sterile worker ants in a colony, doing stereotypic jobs and living out their lives in a relatively noncompetitive ("Marxist") environment. But brain cells—I now think—must compete vigorously in a marketplace. For what?

What could a neuron "want"? The energy and raw materials it needs to thrive—just like its unicellular eukaryote ancestors and more distant cousins, the bacteria and archaea. Neurons are robots; they are certainly not conscious in any rich sense—remember, they are eukaryotic cells, akin to yeast cells or fungi. If individual neurons are conscious, then so is athlete's foot. But neurons are, like these mindless but intentional cousins, highly competent agents in a life-or-death struggle, not in the environment between your toes but in the demanding environment of the brain, where the victories go to those cells that can network more effectively, contribute to more influential trends at the virtual machine levels where large-scale human purposes and urges are discernible.

I now think, then, that the opponent-process dynamics of emotions, and the roles they play in controlling our minds, is underpinned by an "economy" of neurochemistry that harnesses the competitive talents of individual neurons. (Note that the idea is that neurons are still good team players within the larger economy, unlike the more radically selfish cancer cells. Recalling François Jacob's dictum that the dream of every cell is to become two cells, neurons vie to stay active and to be influential, but do not dream of multiplying.)

Intelligent control of an animal's behavior is still a computational process, but the neurons are "selfish neurons," as Sebastian Seung has said, striving to maximize their intake of the different currencies of reward we have found in the brain. And what do neurons "buy" with their dopamine, their serotonin or oxytocin, and so on? Greater influence in the networks in which they participate.

Psychologist and skeptic; author of Conversations on Consciousness

"Where Are You, Sue?"

Imagine me, if you will, in the Oxford of 1970: a new undergraduate, thrilled by the intellectual atmosphere, the hippie clothes, joss-stick-filled rooms, late nights, early morning lectures, and mind-opening cannabis.

I joined the Society for Psychical Research and became fascinated with occultism, mediumship, and the paranormal—ideas that clashed tantalizingly with the physiology and psychology I was studying. Then late one night, something very strange happened. I was sitting around with friends, smoking, listening to music, and enjoying the vivid imagery of rushing down a dark tunnel toward a bright light, when my friend spoke.

I couldn't reply.

"Where are you, Sue?" he asked, and suddenly I seemed to be on the ceiling looking down. "Astral projection!" I thought, and then I (or some imagined flying "I") set off across Oxford, over the country, and way beyond. For more than two hours I fell through strange scenes and mystical states, losing space and time and ultimately my self. It was an extraordinary and life-changing experience. Everything seemed brighter, more real, and more meaningful than anything in ordinary life, and I longed to understand it.

But I jumped to all the wrong conclusions. Perhaps understandably, I assumed that my spirit had left my body and that this proved all manner of things—life after death, telepathy, clairvoyance, and much, much more. I decided, with splendid, youthful overconfidence, to become a parapsychologist and prove all my closed-minded science lecturers wrong. I found a PhD place, funded myself by teaching, and began to test my memory theory of ESP. And this is where my change of mind—and heart, and everything else—came about.

I did the experiments. I tested telepathy, precognition, and clairvoyance; I got only chance results. I trained fellow students in imagery techniques and tested them again; chance results. I tested twins in pairs; chance results. I worked in play groups and nursery schools with very young children (their naturally telepathic minds are not yet warped by education, you see); chance results. I trained as a Tarot reader and tested the readings; chance results.

Occasionally I got a significant result. Oh, the excitement! I responded as I think any scientist should, by checking for errors, recalculating the statistics, and repeating the experiments. But every time I either found the error responsible, or failed to repeat the results. When my enthusiasm waned, or I began to doubt my original beliefs, there was always another corner to turn—always someone saying, "But you must try xxx." It was probably three or four years before I ran out of xxx's.

I remember the very moment when something snapped (or should I say, "I seem to remember . . ." in case it's a false flashbulb memory). I was lying in the bath trying to fit my latest null results into paranormal theory, when it occurred to me for the very first time that I might have been completely wrong, and my tutors right. Perhaps there were no paranormal phenomena at all.

As far as I can remember, this scary thought took some time to sink in. I did more experiments and got more chance results. Parapsychologists called me a "psi-inhibitory experimenter," meaning that I didn't get paranormal results because I didn't believe strongly enough. I studied other people's results and found more errors, and even outright fraud. By the time my PhD was completed, I had become a skeptic.

Until then, my whole identity had been bound up with the paranormal. I had shunned a sensible PhD place and ruined my chances of a career in academia (as my tutor at Oxford liked to say). I had hunted ghosts and poltergeists, trained as a witch, attended spiritualist churches, and stared into crystal balls. But all of that had to go.

Once the decision was made, it was actually quite easy. Like many big changes in life, this one was terrifying in prospect but easy in retrospect. I soon became "rent-a-skeptic," appearing on TV shows to explain how the illusions work, why there is no telepathy, and how to explain near-death experiences by events in the brain.

What remains is a kind of openness to evidence. However firmly I believe in some theory (on consciousness, memes, or whatever), however closely I might be identified with some position or claim, I know that the world won't fall apart if I have to change my mind.

Psychologist, London School of Economics; author of Seeing Red: A Study in Consciousness

Solving the Hard Problem

The economist John Maynard Keynes, when criticized for shifting his position on monetary policy, retorted, "When the facts change, I change my mind. What do you do, sir?" Point taken. Yet, despite the way the *Edge* question has been framed, in science it is not always true that it requires new facts to change people's minds. Instead, as Thomas Kuhn recognized, at major turning points in the history of science, theorists who have previously found themselves struggling to make sense of "known facts" sometimes undergo a radical change in perspective, such that they see these same facts in a quite different light. Where people earlier saw the rabbit, they now see the duck.

In my own research on consciousness, I have changed my mind more than once. I expect it will happen again. But it has not—at least, so far—been because I learned any new facts. Contrary to the hopes of neuroscientists on one side, quantum physicists on the other, I'm pretty sure all the facts we need to solve the hard problem are already familiar to us, if only we could see them for what they are. No magic bullet is going to emerge from the lab, from brain-imaging or particle accelerators. Instead, what we are waiting for is merely (!) a revolutionary new way of thinking about things that we all, as conscious creatures, already know—perhaps a way of making those same facts unfamiliar.

What is the hard problem? The problem is to explain the mysterious out-of-this-world qualities of conscious experience— the felt redness of red, the felt sharpness of pain. I once believed that the answer lay in introspection: "Thoughts about thoughts," I reckoned, could yield the requisite magical properties as an emergent property. But I later realized on logical (not factual) grounds that this idea was empty. Magic doesn't simply emerge, it has to be constructed. So, since then, I've been working on a constructivist theory of consciousness. And my most promising line yet (as I see it) has been to turn the problem round and to imagine that the hardness of the problem may actually be the key to its solution.

Just suppose that the Cartesian theater of consciousness, about which modern philosophers are generally so skeptical, is in fact a biological reality. Suppose, indeed, that nature has designed our brains to contain a mental theater, designed for the very purpose of staging the qualia-rich spectacle on which we set such store. Suppose, in short, that consciousness exists primarily for our entertainment and amazement.

I may tell you that, with this changed mindset, I already see the facts quite differently. I trust that—when you catch on—it will do the same for you, madam (and sir).

BARRY C. SMITH

Philosopher, Birkbeck School of Philosophy, University of London; coeditor (with Crispin Wright and Cynthia Macdonald) of Knowing Our Own Minds

Neuroscience and Philosophy

For a long time I regarded neuroscience as a fascinating source of information about the workings of the visual system and its dual pathways for sight and action about the fear system in humans and animals, and about numerous puzzling pathology cases arising from site-specific lesions.

Yet, despite the interest of these findings, I had little faith that the profusion of fMRI studies of different cortical regions would tell us much about the problems that had preoccupied philosophers for centuries. After all, some of the greatest minds of history had long pondered the nature of consciousness, the self, the relation between self and others, only to produce a greater realization of how hard it was to say something illuminating about any of these phenomena. The more one is immersed in neural mechanisms, the less one seems to be talking about consciousness, and the more one attends to the qualities of conscious experience, the less easy it is to connect with the mechanism of the brain. In despair, some philosophers suggested that we must reduce or eliminate the everyday way of speaking about our mental lives to arrive at a science of mind. There appeared to be a growing gulf between how things appeared to us and how reductionist neuroscience told us they were.

However, I have changed my mind about the relevance of neuroscience to philosophers' questions, and vice versa. Why? Well, first, because the most interesting findings in cognitive neuroscience are not in the least reductionist. On the contrary, neuroscientists rely on subjects' reports of their experiences in familiar terms to target the states they wish to correlate with increased activity in the cortex. Researchers disrupt specific cortical areas with transcranial magnetic stimulation (TMS) to discover how subjects' experiences or cognitive capacities are altered.

This search for the neural correlates of specific states and abilities has proved far more successful than any reductionist program, the aim being to explain precisely which neural areas are responsible for sustaining the experiences we typically have as human subjects. And what we are discovering is just how many subsystems cooperate to maintain a unified and coherent field of conscious experience in us. When any of these systems is damaged, what results are bizarre pathologies of mind we find it hard to comprehend. It is here that neuroscientists seek the help of philosophers in analyzing the character of normal experience and describing the nature of the altered states. Reciprocally, what philosophers are learning from neuroscience is leading to revisions in cherished philosophical views, mostly for the better. For example, the early stages of sensory processing show considerable cross-modal influence of one sense on another: The nose smells what the eye sees, the tongue tastes what the ear hears, the recognition of voice is enhanced by, and enhances, facial recognition in the fusiform face area—all of which leads us to conclude that the five senses are not nearly as separate as common sense and most philosophers have always assumed.

Similar breakthroughs in understanding how our sense of self depends on the somatosensory system are leading to revised philosophical thinking about the nature of self. And while philosophers have wondered how individuals come to know about the minds of others, neuroscience assumes the problem to have been partly solved by the discovery of the mirror-neuron system, which suggests an elementary, almost bodily, level of intersubjective connection between individuals, from which the more sophisticated notions of self and other may develop. We don't start, like Descartes, with the self and bridge to our knowledge of other minds. We start instead with primitive social interactions, from which the notions of self and other are constructed.

Neuroscientists present us with strange phenomena, such as patients with lesions in the right parietal region who are convinced that their left arm does not belong to them. Some still feel sensations of pain in that hand but do not believe that it is *their* pain that is felt—something philosophers previously believed to be conceptually impossible.

I think the startling conclusion should be just how precarious the typical experience of the normally functioning mind really is. We should not find it strange to come across people who do not believe their hand belongs to them, or that it acts under someone else's command. Instead, we should think how remarkable it is that this assembly of subsystems that keeps track of our limbs, our volitions, our position in space, and our recognition of others should cooperate to sustain the sense of self and the feeling of a coherent and unified experience of the world so familiar to us that philosophers have believed it to be the most certain thing we know. It isn't the pathology cases of cognitive neuropsychology that are exceptional, it is the normally functioning minds that we should find the most surprising.

JESSE BERING

Director of the Institute of Cognition and Culture, Queen's University, Belfast

Wiggle Room

If asked years ago whether I believed in God, my answer would have gone something like this: "I believe there's something. . . ." This response leaves enough wiggle room for a few quasi-religious notions to slip comfortably through. I no longer believe that my soul is immortal, that the universe sends me messages every now and then, or that my life story will unfold according to some inscrutable plan. But it is more like knowing how and why a perceptual illusion is deceiving my evolved senses than it is becoming immune to the illusion altogether.

Here's a snapshot of how these particular illusions work:

Psychological Immortality: There's a scene in André Gide's *The Counterfeiters* in which a suicidal man puts a pistol to his temple but hesitates for fear of the noise from the blast. Similarly, a group of college students who rejected the idea that consciousness survives death nonetheless told me that someone who had died in a car accident would know he was dead. "There is no afterlife," one participant said. "He sees that now."

In wondering what it's like to be dead, our psychology responds by running mental simulations using previous states of consciousness. The trouble is that death is not like anything we've ever experienced—or can experience. (What is it like to be conscious yet unconscious at the same time?) I doubt you'd find anyone who believes less in the afterlife, yet I have a very real fear of ghosts and I feel guilty for not visiting my mother's grave more often.

Symbolic Meaning of Natural Events: Psychologist Becky Parker and I told a seven-year-old that an invisible princess was in the room with her. The task was to find a hidden ball by placing her hand on top of the box she thought it was inside. If you change your mind, we said, just move your hand to the other box. Now, Princess Alice likes you, and she's going to help you find the ball. "I don't know how she's going to tell you," said Becky, "but somehow she'll tell you if you pick the wrong box."

The child picked a box, held her hand there, and after fifteen seconds the box opened to reveal the ball (there were balls in both boxes). On the second trial, as soon as the child chose a box, a picture crashed to the ground and the child moved her hand to the other box. In doing so, she responded just like most other seven-year-olds we tested. They didn't need to believe in Princess Alice to see the picture falling as a sign. In fact, if skepticism can be operationally measured by the degree of tilt in rolling eyeballs, many of them could be called skeptics.

More surprising was that slightly younger children, the credulous five-year-olds, didn't move their hands when the picture fell, and when asked why it fell they said things like, "I don't know why she did it, she just did it." They saw Princess Alice as running about making things happen, not as a communicative partner. To them, the events had nothing to do with their own behavior. Finally, the three-year-olds we tested simply shrugged their shoulders and said that the picture was broken. Princess Alice who?

Seeing signs in natural events is a developmental accomplishment, rather than the result of a gap in scientific knowledge. To experience an illusion, the psychological infrastructure must first be in place. Whenever I hear mayors blaming hurricanes on drug use or evangelicals attributing tsunamis to homosexuality, I think of Princess Alice. Still, after receiving bad news my first impulse is to ask myself, "Why me?" Even for someone like me, scientific explanations just don't scratch the itch like supernatural ones.

Personal Destiny: Jean-Paul Sartre, the atheistic existentialist, observed that he couldn't help but feel as though a divine hand had guided his life. "It contradicts many of my other ideas," he said. "But it is there, floating vaguely. And when I think of myself I often think rather in this way, for want of being able to think otherwise."

My own atheism is not as organic as was Sartre's. Only scientific evidence and eternal vigilance have enabled me to step outside of this particular illusion of personal destiny. Psychologists now know that human beings intuitively reason as though natural categories exist for an intelligently designed purpose. Clouds don't just exist, say kindergartners, they're there for raining.

Erring this way about clouds is one thing, but when it colors our reasoning about our own existence, that's where this teleofunctional bias gets really interesting. The illusion of personal destiny is intricately woven with other quasi-religious illusions in a complex web that researchers have not even begun to pull apart. My own private thoughts remain curiously saturated with doubts about whether I'm doing what I'm "meant" for.

Some beliefs are arrived at so easily, held so deeply, and divorced so painfully that it seems unnatural to give them up. Such beliefs can be abandoned when the illusions giving rise to them are punctured by scientific knowledge, but a mind designed by nature cannot be changed fundamentally. I stopped believing in God long ago, but he still casts a long shadow.

President of the Royal Society; professor of cosmology and astrophysics, master, Trinity College, Cambridge; author of Our Final Hour: A Scientist's Warning

We Are Custodians
of a Posthuman Future

Public discourse on very long-term planning is riddled with inconsistencies. Mostly we discount the future very heavily. Investment decisions are expected to pay off within a decade or two, but when we look further ahead—in discussions of energy policy, global warming, and so forth—we underestimate the possible pace of transformational change. In particular, we need to keep our minds open, or at least ajar, to the possibility that humans themselves could change drastically within a few centuries.

Our medieval forebears in Europe had a cosmic perspective that was a millionfold more constricted than ours. Their entire cosmology, from creation to apocalypse, spanned only a few thousand years. Today, the stupendous time spans of the evolutionary past are part of common culture, except among some creationists and fundamentalists. Moreover, we are mindful of immense future potential. It seems absurd to regard humans as

the culmination of the evolutionary tree. Any creatures witness-
ing the sun's demise six billion years hence won't be human;
they could be as different from us as we are from slime mold.

But despite these hugely stretched conceptual horizons, the
timescale on which we can sensibly plan or make confident forecasts has got shorter rather than longer. Medieval people, despite
their constricted cosmology, did not expect drastic changes
within a human life; they devotedly added bricks to cathedrals
that would take a century to finish. For us, unlike for them, the
next century will surely be drastically different from the present.
There is a huge disjunction between the ever-shortening time-
scales of historical and technical change and the near-infinite
time spans over which the cosmos itself evolves.

Human-induced changes are occurring with runaway speed.
It's hard to predict a mere century from now, because what
will happen depends on us. This is the first century in which
humans can collectively transform, even ravage, the entire biosphere. Humanity will soon itself be malleable, to an extent that
is qualitatively new in the history of our species. New drugs (and
perhaps even brain implants) could change human character;
the cyberworld has potential that is both exhilarating and fright-
ening. We cannot confidently foresee lifestyles, attitudes, social
structures, or population sizes a century hence. Indeed, it's not
even clear for how long our descendants will remain distinctively
"human." Darwin himself noted that "not one living species will
transmit its unaltered likeness to a distant futurity." Our species
will surely change and diversify faster than any predecessor, via
human-induced modifications—whether intelligently controlled
or unintended—not by natural selection alone. Just how fast this
could happen is disputed by experts, but the posthuman era may
be only centuries away.

These thoughts might seem irrelevant to practical discussions and best left to speculative academics and cosmologists.

I used to think this. But humans are now, individually and collectively, so greatly empowered by rapidly changing technology that we can, by design or as unintended consequences, engender global changes that resonate for centuries. And, sometimes at least, policymakers indeed think far ahead.

The global warming induced by fossil fuels burnt in the next fifty years could trigger gradual sea-level rises that continue for a millennium or more. And in assessing sites for radioactive waste disposal, governments impose the requirements that they be secure for ten thousand years.

It's real political progress that these long-term challenges are higher on the international agenda and that planners seriously worry about what might happen more than a century hence. But in such planning we need to be mindful that it may not be people like us who confront the consequences of our actions today. We are custodians of a "posthuman future"—here on Earth and perhaps beyond—that cannot just be left to writers of science fiction.

JANNA LEVIN

Physicist, Columbia University; author of A Madman Dreams
of Turing Machines

Finite and Edgeless

There are innumerable little things about which I've changed my mind, but the size of the universe is literally the biggest physical attribute that has inspired a radical change in my thinking. I won't claim I "believe" the universe is finite, just that I recognize that a finite universe is a realistic possibility for our cosmos.

The general theory of relativity describes local curves in spacetime due to matter and energy. This model of gravity as a warped spacetime has seen countless successes, beginning with the confirmation of an anomaly in the orbit of Mercury and continuing with the predictions of the existence of black holes, the expansion of spacetime, and the creation of the universe in a Big Bang. However, general relativity says very little about the global shape and size of the universe. Two spaces can have the same curvature locally but very different global properties. A flat space, for instance, can be infinite, but there is another possibility—that it is finite and edgeless, wrapped back onto itself like a doughnut but still flat. And there are an infinite number of ways of folding spacetime into finite, edgeless shapes, a kind of cosmic origami.

I grew up believing the universe was infinite. It was never taught to me, in the sense that no one ever tried to prove to me that the universe was infinite. It just seemed a natural assumption based on simplicity. That sense of simplicity no longer resonates as true once we have confronted the idea that there must be a theory of gravity beyond general relativity that involves the quantization of spacetime. In cosmology, we have become accustomed to models of the universe that invoke extra dimensions, all of which are finite, and it seems fair to imagine a universe born with all of its dimensions finite and compact. Then we are left with the mystery of why only three dimensions become so incredibly huge, while the others remain curled up and small. We even hope to test models of extra dimensions in imminent laboratory experiments. These ideas are not remote and fantastical; they are testable.

People have told me they were surprised (disappointed) to hear me suggest the universe was finite. The infinite universe, they believe, is full of infinite potential and so, philosophically (emotionally), so much richer and more thrilling. I explain that my suggestion of a finite universe is not a moral failing on my part or a consequence of diminished imagination. More thrilling is the knowledge that it does not matter what I believe. It does not matter whether I prefer an infinite universe or a finite universe. Nature is not designed to satisfy our personal longings. Nature is what she is, and it's a privilege merely to be privy to her mathematical codes.

I don't know that the universe is finite, and so I don't believe that it is finite. I do see, however, that our mathematical reasoning has led to remarkable and sometimes psychologically uncomfortable discoveries. And I do believe that it is a realistic possibility that one day we may discover the shape of the universe. If the universe is too vast for us to ever observe the extent of space, we may still discover the size and shape of internal dimensions. From small extra dimensions, we might possibly be able to infer the size and shape of the large dimensions. Until then, I won't make up my mind.

JOSEPH LEDOUX

Neuroscientist, New York University; author of The Synaptic Self

Reconsolidating Memory

Like many scientists in the field of memory, I used to think that a memory is something stored once in the brain and accessed when used.

Then, in 2000, Karim Nader, a researcher in my lab, did an experiment that convinced me and many others that our usual way of thinking was wrong. What he showed was that each time a memory is used, it has to be restored as a new memory in order to be accessible later. The old memory is either not there or is inaccessible.

In short, your memory of something is only as good as your last memory of it. That is why people who witness crimes testify about what they read in the paper rather than what they witnessed. Research on this topic, called reconsolidation, has become the basis of a possible treatment for post-traumatic stress disorder, drug addiction, and any other disorder based on learning.

How much did I change my mind? When Karim proposed to do the study, I told him it was a waste of time. His results had a profound impact on me. I'm not swayed by arguments based on faith. I can be moved by good logic. But I am always swayed by a good experiment, even if it goes against my cherished beliefs. I might not give up on a scientific belief after one experiment, but when the evidence mounts over multiple studies, I change my mind.

NICHOLAS CARR

Writer on business, technology, and culture; author of The Big Switch: Rewiring the World, from Edison to Google

The Internet and the Centralization of Power

In January 2007, China's president, Hu Jintao, gave a speech before a group of Communist Party officials. His subject was the Internet. "Strengthening network culture construction and management," he assured the assembled bureaucrats, "will help extend the battlefront of propaganda and ideological work. It is good for increasing the radiant power and infectiousness of socialist spiritual growth."

If I had read those words a few years earlier, they would have struck me as ludicrous. It seemed so obvious that the Internet stood in opposition to the kind of centralized power symbolized by China's regime. A vast array of autonomous nodes, not just decentralized but centerless, the Net was a technology of personal liberation, a force for freedom.

I now see that I was naive. Like many others, I mistakenly interpreted a technical structure as a metaphor for human liberty. In recent years, we have seen clear signs that while the Net

37

may be a decentralized communications system, its technical and commercial workings actually promote the centralization of power and control. Look, for instance, at the growing concentration of Web traffic. From 2002 through 2006, the number of Internet sites nearly doubled, yet the concentration of traffic at the ten most popular sites nonetheless grew substantially, from 31 percent to 40 percent of all page views, according to the research firm Compete.

Or look at how Google continues to expand its hegemony over Web searching. In March 2006, the company's search engine was used to process a whopping 58 percent of all searches in the United States, according to Hitwise. By November 2007, the figure had increased to 65 percent. The results of searches are also becoming more, not less, homogeneous. Do a search for any common subject and you're almost guaranteed to find Wikipedia at or near the top of the list of results.

It's not hard to understand how the Net promotes centralization. For one thing, its prevailing navigational aids, such as search-engine algorithms, form feedback loops. By directing people to the most popular sites, they make those sites even more popular. On the Web, as elsewhere, people stream down the paths of least resistance.

The predominant means of making money on the Net— collecting small sums from small transactions—also promotes centralization. It is only by aggregating vast quantities of content, data, and traffic that businesses can turn large profits. That's why companies like Microsoft and Google have been so aggressive in buying up smaller Web properties. Google, which has been acquiring companies at the rate of about one a week, has disclosed that its ultimate goal is to "store 100 percent of user data."

As the dominant Web companies grow, they are able to gain ever-larger economies of scale through massive capital investments in the "server farms" that store and process online data. That, too, promotes consolidation and centralization. Executives of Yahoo! and Sun Microsystems have recently predicted that control over the Net's computing infrastructure will ultimately lie in the hands of five or six organizations.

To what end will the Web giants deploy their power? They will, of course, seek to further their own commercial or political interests by monitoring, analyzing, and manipulating the behavior of "users." The connection of previously untethered computers into a single programmable system has created "a new apparatus of control," to quote NYU's Alexander Galloway. Even though the Internet has no center, technically speaking, control can be wielded, through software code, from anywhere. What's different, in comparison to the physical world, is that acts of control are more difficult to detect.

So it's not Hu Jintao who is deluded in believing that the Net might serve as a powerful tool for central control, it is those who assume otherwise. I used to count myself among them. But I've changed my mind.

DOUGLAS RUSHKOFF

Media analyst; documentary writer; author of Get Back in the Box: Innovation from the Inside Out

Cyberspace Has Become Just Another Place to Do Business

I thought the Internet would change people. I thought it would allow us to build a new world through which we could model new behaviors, values, and relationships. In the '90s, I thought the experience of going online for the first time would change a person's consciousness as much as if he or she had dropped acid in the '60s.

I thought Amazon.com was a ridiculous idea, and that the Internet would shrug off business as easily as it did its original Defense Department minders.

For now, at least, it's turned out to be different.

Virtual worlds like Second Life have been reduced to market opportunities: Advertisers from banks to soft drinks purchase space and create fake characters, while kids (and Chinese digital sweatshop laborers) earn "play money" in the game only to sell it to lazier players on eBay for real cash.

The businesspeople running Facebook and MySpace are rivaled only by the members of these online "communities" in their willingness to surrender their identities and ideals for a buck, a click-through, or a better market valuation.

The open-source ethos has been reinterpreted through the lens of corporatism as "crowd sourcing"—meaning just another way to get people to do work for no compensation. And even file sharing has been reduced to a frenzy of acquisition that has less to do with music than with the ever-expanding hard drives of successive iPods.

Sadly, cyberspace has become just another place to do business. The question is no longer how browsing the Internet changes the way we look at the world; it's which browser we'll be using to buy and sell stuff in the same old world.

MARCELO GLEISER

Physicist, Dartmouth College; author of The Prophet and
the Astronomer

To Unify or Not to Unify?

I grew up infused with the idea of unification. It came first
from religion, from my Jewish background. God was all over, was
all-powerful, and had a knack for interfering with human affairs,
at least in the Old Testament. He then appeared to have decided
to be a bit shyer, sending a Son instead and revealing Himself
only through visions and prophecies. Needless to say, when, as
a teenager, I started to get interested in science, this concept of
an all-pervading God, these stories of floods, commandments,
and plagues, became very suspect. I turned to physics, idolizing
Einstein and his science; here was a Jew who saw further, who
found a way of translating the old monotheistic tradition into the
universal language of science.

As I began my research career, I had no doubt that I wanted
to become a theoretical physicist working on particle physics and
cosmology. Why the choice? Simple: It was the joining of the
two worlds of the very large and the very small that offered the
best hope for finding a unified theory of all nature—that would
bring together matter and forces into one single magnificent for-

mulation, the final Platonist triumph. This was what Einstein tried to do for the last three decades of his life, although in his time it was more a search for unifying only half of the forces of nature: gravity and electromagnetism.

I wrote dozens of papers related to the subject of unification; even my PhD dissertation was on the topic. I was fascinated by the modern approaches to the idea—supersymmetry, superstrings, a space with extra, hidden dimensions. A part of me still is. But then, a few years ago, something snapped.

It probably was brought about by a combination of factors—a deeper understanding of the historical and cultural processes that shape scientific ideas. I started to doubt unification, finding it to be the scientific equivalent of a monotheistic formulation of reality, a search for God revealed in equations. Of course, had we the slightest experimental evidence in favor of unification—of supersymmetry and superstrings—I would have been the first to pop the champagne open. But it's been over twenty years and all attempts so far have failed. Nothing in particle accelerators, nothing in cryogenic dark-matter detectors, no magnetic monopoles, no proton decay—all those tell-tale signs of unification predicted over the years. Even our wonderful Standard Model of Particle Physics, by which we formulate the unification of electromagnetism and the weak nuclear interactions, is not really a true unification: The theory retains information from both interactions in the form of their strengths—or, in more technical jargon, of their coupling constants. A true unification should have a single coupling constant, a single interaction.

All my recent antiunification convictions can crumble during the next few years, after our big new machine, the Large Hadron Collider at CERN, is turned on. Many colleagues hope that supersymmetry will finally show its face. Others even bet on possible signs of extra dimensions. However, I have a feeling things won't turn out so nicely. The model of unification, so aes-

thetically appealing, may be only that—an aesthetically appealing description of nature that, unfortunately, doesn't correspond
to physical reality. Nature doesn't share our myths. The stakes
are high indeed, but, being a mild agnostic, I don't believe until
there is evidence. And then there is no need to believe any longer, which is precisely the beauty of science.

FREEMAN DYSON

Physicist, Institute for Advanced Study, Princeton University; author of A Many-Colored Glass: Reflections on the Place of Life in the Universe

Demolishing Myths

When facts change your mind, that's not always science. It may be history. I changed my mind about an important historical question: Did the nuclear bombings of Hiroshima and Nagasaki bring World War II to an end? Until this year I used to say, Perhaps. Now, because of new facts, I say no. This question is important, because the myth of the nuclear bombs bringing the war to an end is widely believed. To demolish this myth may be a useful first step toward ridding the world of nuclear weapons.

Until the last few years, the best summary of evidence concerning this question was a book, *Japan's Decision to Surrender*, by Robert Butow, published in 1954. Butow interviewed the surviving Japanese leaders who had been directly involved in the decision. He asked them whether Japan would have surrendered if the nuclear bombs had not been dropped. His conclusion: "The Japanese leaders themselves do not know the answer to that question, and if they cannot answer it, neither can I." Until recently, I believed what the Japanese leaders said

45

to Butow, and I concluded that the answer to the question was unknowable.

Facts causing me to change my mind were brought to my attention by Ward Wilson. Wilson summarized the facts in an article, "The Winning Weapon? Rethinking Nuclear Weapons in Light of Hiroshima," in the spring 2007 issue of the magazine *International Security*. He gives references to primary source documents and to analyses published by other historians, in particular by Robert Pape and Tsuyoshi Hasegawa. The facts are as follows:

1. Members of the Supreme Council, which customarily met with the Emperor to take important decisions, learned of the nuclear bombing of Hiroshima on the morning of August 6, 1945. Although Foreign Minister Togo asked for a meeting, no meeting was held for three days.

2. A surviving diary records a conversation of Navy Minister Yonai, who was a member of the Supreme Council, with his deputy on August 8. The Hiroshima bombing is mentioned only incidentally. More attention is given to the fact that the rice ration in Tokyo is to be reduced by 10 percent.

3. On the morning of August 9, Soviet troops invaded Manchuria. Six hours after hearing this news, the Supreme Council was in session. News of the Nagasaki bombing, which happened the same morning, reached the Council only after the session started.

4. The August 9 session of the Supreme Council resulted in the decision to surrender.

5. The Emperor, in his rescript to the military forces ordering their surrender, does not mention the nuclear bombs but emphasizes the historical analogy between

the situation in 1945 and the situation at the end of the Sino-Japanese war in 1895. In 1895 Japan had defeated China, but accepted a humiliating peace when European powers, led by Russia, moved into Manchuria and the Russians occupied Port Arthur. By making peace, the Emperor Meiji had kept the Russians out of Japan. Emperor Hirohito had this analogy in his mind when he ordered the surrender.

6. The Japanese leaders had two good reasons for lying when they spoke to Robert Butow. The first reason was explained afterward by Lord Privy Seal Kido, another member of the Supreme Council: "If military leaders could convince themselves that they were defeated by the power of science but not by lack of spiritual power or by strategic errors, they could save face to some extent." The second reason was that they were telling the Americans what the Americans wanted to hear, and the Americans did not want to hear that the Soviet invasion of Manchuria brought the war to an end.

In addition to the myth of two nuclear bombs bringing the war to an end, there are other myths that need to be demolished. There is the myth that, if Hitler had acquired nuclear weapons before we did, he could have used them to conquer the world. There is the myth that the invention of the hydrogen bomb changed the nature of nuclear warfare. There is the myth that international agreements to abolish weapons without perfect verification are worthless. All these myths are false. After they are demolished, dramatic moves toward a world without nuclear weapons may become possible.

ROGER HIGHFIELD

Science editor of New Scientist; *coauthor (with Ian Wilmut) of*
After Dolly: The Uses and Misuses of Cloning

Unfettered by Facts

I am a heretic. I have come to question the key assumption behind this *Edge* survey: "When facts change your mind, that's science." This idea that science is an objective fact-driven pursuit is laudable, seductive, and—alas—a mirage.

Science is a never-ending dialogue between theorists and experimenters. But people are central to that dialogue. And people ignore facts. They distort them or select the ones that suit their cause, depending on how they interpret their meaning. Or they don't ask the right questions to obtain the relevant facts.

Contrary to the myth of the ugly fact that can topple a beautiful theory—and against the grain of our lofty expectations—scientists sometimes play fast and loose with data, highlighting what suits them and ignoring stuff that doesn't.

The harsh spotlight of the media often encourages them to strike a confident pose even if the facts don't. I am often struck by how facts are ignored, insufficient, even abused. I back well-designed animal research, but I'm puzzled by scientists' ignoring the basic fact that vivisection is inefficient at generating cures for human disease. Intelligent design is for cretins, but despite the endless proselytizing about the success of Darwin—assuming that evolution is a fact—I could still see that success as being superseded, rather as Einstein's ideas replaced Newton's law of gravity. I believe in man-made global warming, but computer-projected facts that purportedly tell us what is in store for Earth in the next century leave me cold.

I support embryo research but was irritated by one oft-cited fact in the recent British debate on the manufacture of animal-human hybrid embryos: "Only a tiny fraction" of the hybrid made by the Dolly cloning method (nuclear transfer) contains animal DNA. Given that it features in mitochondria, which are central to a range of diseases; given that a single spelling mistake in DNA can be catastrophic; and given that no one really understands what nuclear transfer does, this "fact" was propaganda.

Some of the most exotic and prestigious parts of modern science are unfettered by facts. I have recently written about whether our very ability to study the heavens may have shortened the inferred lifetime of the cosmos, whether there are two dimensions of time, even the prospect that time itself could cease to be in billions of years. The field of cosmology is in desperate need of more facts, as highlighted by the aphorisms made at its expense. ("There is speculation, pure speculation, and cosmology." "Cosmologists are often in error, never in doubt.")

Scientists have to make judgments about the merits of new facts. Ignoring them in the light of strong intuition is the mark of a great scientist. Take Einstein, for example: When Walter Kaufmann claimed to have experimental facts that refuted special relativity, Einstein stuck to his guns and was proved right. Equally, Einstein's intuition misled him in his last three decades, when he pursued a fruitless quest for a unified field theory that was not helped by his lack of interest in novel facts: the new theoretical ideas, particles, and interactions that had emerged at this time.

When it comes to work in progress, in particular, many scientists treat science like a religion: The facts should be made to fit the creed. However, facts are necessary for science but not sufficient. Science is when, in the face of extreme skepticism, enough facts accrue to change lots of minds.

Our rising and now excessive faith in facts alone can be seen in a change in the translation of the motto of the world's oldest academy of science, the Royal Society. *Nullius in Verba* was once taken to mean "On the word of no one," to highlight the extraordinary power that empirical evidence bestowed upon science. The message was that experimental evidence trumped personal authority.

Today the Society talks of the need to "verify all statements by an appeal to facts determined by experiment." But whose facts? Was it a well-designed experiment? And are we getting all the relevant facts? The Society should adopt the snappier version that captures its original spirit: "Take nobody's word for it."

DANIEL ENGBER

Science editor of Slate *magazine*

Animal Sacrifice

Two years ago, I watched the Dalai Lama address thousands of laboratory biologists at the Society for Neuroscience meeting in Washington, D.C. At the end of his speech, someone asked about the use of animals in lab research. "That's difficult," replied His Holiness. "Always stress the importance of compassion. . . . In highly necessary experiments, try to minimize pain."

The first two words of his answer provided most of the moral insight.

Universities already have cumbersome animal-research protocols in place to eliminate unnecessary suffering, and few lab workers would do anything but try to minimize pain.

When I first entered graduate school, this Western-cum-Buddhist policy seemed like a neat compromise between protecting animals and supporting the advance of knowledge. But after I'd spent several years cutting up mice, birds, kittens, and monkeys, my mind was changed.

Not because I was any less dedicated to the notion of animal research: I still believe it's necessary to sacrifice living things in the name of scientific progress. But I saw how institutional safeguards served to offload the moral burden from the researchers themselves.

Rank-and-file biologists are rarely asked to consider the key ethical questions around which these policies are based. True, the National Institutes of Health has for almost twenty years required that graduate training institutions offer a course in responsible research conduct. But in the class I took, we received PR advice rather than moral guidance: What's the best way to keep your animal research out of the public eye?

In practice, I found that scientists were far from monolithic in their attitudes toward animal work. (*Drosophila* researchers had misgivings about the lab across the hall, where technicians perfused the still-beating hearts of mice with chemical fixative; mouse researchers didn't want to implant titanium posts in the skulls of water-starved monkeys.) They weren't animal-rights zealots, of course; they had nothing but contempt for the PETA protestors who passed out fliers in front of the lab buildings. But they did have real misgivings about the extent to which biology research might go in its exploitation of living things.

At the same time, very few of us took the time to consider whether or how we might sacrifice fewer animals (or no animals at all). Why bother, when the Institutional Animal Care and Use Committee had already signed off on the research? The hard part of this work isn't convincing an IACUC board to sanction the killing, it's making sure you've exhausted every possible alternative.

RAY KURZWEIL

Inventor and technologist; author of The Singularity Is Near:
When Humans Transcend Biology

Where Are They?

I've come to reject the common SETI (search for extraterrestrial intelligence) wisdom that there must be millions of
technology-capable civilizations within our "light sphere" (the
region of the universe accessible to us by electromagnetic communication). The Drake formula provides a means to estimate
the number of intelligent civilizations in a galaxy or in the universe. Essentially, the likelihood of a planet evolving biological
life that has created sophisticated technology is tiny, but there
are so many star systems that there should still be many millions
of such civilizations. Carl Sagan's analysis of the Drake formula
concluded that there should be around a million civilizations
with advanced technology in our galaxy, while Frank Drake
himself estimated around ten thousand. And there are many
billions of galaxies. Yet we don't notice any of these intelligent
civilizations, hence the paradox that Fermi raised in his famous
comment: "Where is everybody?"

We can readily explain why any one of these civilizations
might be quiet. Perhaps it destroyed itself. Perhaps it is following

53

the *Star Trek* ethical guideline to avoid interference with primitive civilizations (such as ours). These explanations make sense for any one civilization, but it is not credible, in my view, that every one of the billions of technology-capable civilizations that should exist has destroyed itself or decided to remain quiet.

The SETI project is sometimes described as trying to find a needle (evidence of a technical civilization) in a haystack (all the natural signals in the universe). But actually, any technologically sophisticated civilization would be generating trillions of trillions of needles (noticeably intelligent signals). Even if they have switched away from electromagnetic transmissions as a primary form of communication, there would still be vast artifacts of electromagnetic phenomena generated by all of the many computational and communication processes that such a civilization would need to engage in.

Now let's factor in what I call the law of accelerating returns (the inherent exponential growth of information technology). The common wisdom (based on what I call the intuitive linear perspective) is that it would take many thousands, if not millions, of years for an early technological civilization to become capable of technology that spanned a solar system. But because of the explosive nature of exponential growth, it will take only a quarter of a millennium (in our own case) to go from sending messages on horseback to saturating the matter and energy in our solar system with sublimely intelligent processes.

The price-performance of computation went from 10^{-5} to 10^8 cps (calculations per second) per one thousand dollars in the twentieth century. We also went from about a million dollars to a trillion dollars in the amount of capital devoted to computation, so overall progress in nonbiological intelligence went from 10^{-2} to 10^{17} cps in the twentieth century, which is still short of the human biological figure of 10^{26} cps. By my calculations, however, we will achieve around 10^{69} cps by the end of the twenty-first century, thereby greatly multiplying the intellectual capability of our human-machine civilization. Even if we find communication methods superior to electromagnetic transmissions, we will nonetheless be generating an enormous number of intelligent electromagnetic signals.

According to most analyses of the Drake equation, there should be billions of civilizations, and a substantial fraction of these should be ahead of us by millions of years. That's enough time for many of them to be capable of vast galaxy-wide technologies. So how can it be that we haven't noticed any of the trillions of trillions of "needles" that each of these billions of advanced civilizations should be creating?

My own conclusion is that they don't exist. If it seems unlikely that we would be in the lead in the universe, here on the third planet of a humble star in an otherwise undistinguished galaxy, it's no more perplexing than the existence of our universe, with its ever so precisely tuned formulas to allow life to evolve in the first place.

Mathematician, computer scientist; CyberPunk pioneer; novelist; author of Postsingular

Might Robots See God?

Studying mathematical logic in the 1970s, I believed it was possible to put together a convincing argument that no computer program can fully emulate a human mind. Although nobody had quite gotten the argument right, I hoped to straighten it out.

My belief in this will-o'-the-wisp was motivated by a gut feeling that people have numinous inner qualities that will not be found in machines. For one thing, our self-awareness lets us reflect on ourselves and get into endless mental regresses: "I know that I know that I know. . . ." For another, we have moments of mystical illumination when we seem to be in contact, if not with God, then with some higher cosmic mind. I felt that surely no machine could be self-aware or experience the divine light.

At that point, I'd never actually touched a computer; they were still inaccessible, stygian tools of the establishment. Three decades rolled by, and I'd morphed into a Silicon Valley computer scientist, in constant contact with nimble chips. Setting aside my old prejudices, I changed my mind—and came to believe that we can in fact create humanlike computer programs.

Although writing out such a program is in some sense beyond the abilities of any one person, we can set up simulated worlds in which such computer programs evolve. Some relatively simple setup will, in time, produce a humanlike program capable of emulating all known intelligent human behaviors—writing books, painting pictures, designing machines, creating scientific theories, discussing philosophy, even falling in love. More than that, we will be able to generate an unlimited number of such programs, each with its own particular style and personality.

What of the old-style attacks from the quarters of mathematical logic? Roughly speaking, these arguments always hinged on a spurious belief that we can somehow discern between humanlike systems that are fully reliable and humanlike systems fated to begin spouting gibberish. But the correct deduction from mathematical logic is that there is absolutely no way to separate the sheep from the goats. Note that this is already our situation vis-à-vis real humans: You have no way to tell if and when a friend or a loved one will forever stop making sense.

With the rise of new practical strategies for creating humanlike programs and the collapse of the old a priori logical arguments against this endeavor, I have to reconsider my former reasons for believing humans to be different from machines. Might robots become self-aware? And—not to put too fine a point on it—might they see God? I believe both answers are yes.

Consciousness probably isn't that big a deal. A simple pair of facing mirrors exhibit a kind of endlessly regressing self-awareness, and this type of pattern can readily be turned into computer code.

And what about basking in the divine light? Certainly if we take a reductionistic view that mystical illumination is just a bath of intoxicating brain chemicals, then there seems to be no reason that machines couldn't occasionally be nudged into exceptional states as well. But I prefer to suppose that mystical experiences involve an objective union with a higher level of mind, possibly mediated by offbeat physics such as quantum entanglement, dark matter, or higher dimensions.

Might a robot enjoy these true mystical experiences? Based on my studies of the essential complexity of simple systems, I feel that any physical object at all must be equally capable of enlightenment. As the Zen apothegm has it, "The universal rain moistens all creatures."

ED REGIS

Science writer; author of What is Life?

Seeing Ahead

I used to think you could predict the future. In *Profiles of the Future*, Arthur C. Clarke made it seem so easy. So did all those other experts who confidently predicted the paperless office, the artificial intelligentsia who for decades predicted "human equivalence in ten years," the nanotechnology prophets who kept foreseeing major advances in molecular manufacturing within fifteen years, and so on.

Mostly, the predictions of science and technology types were wonderful: space colonies, flying cars in everyone's garage, the conquest (or even reversal) of aging. There were, of course, the doomsayers, too, such as the population-bomb theorists who said the world would run out of food by the turn of the century.

But at last, after watching all those forecasts not come true — and in fact be falsified in a crashing, breathtaking manner — I began to question the entire business of making predictions. If even Nobel Prize–winning scientists, such as Ernest Rutherford, who gave us essentially the modern concept of the nuclear atom, could say, as he did in 1933, that "We cannot control atomic energy to an extent which would be of any value commercially, and I believe we are not likely ever to be able to do so," what hope was there for the rest of us?

I finally decided I knew the source of this incredible mismatch between confident forecast and actual result. The universe is a complex system, in which countless causal chains act and interact independently and simultaneously, the ultimate nature of some of them unknown to science even today. There are in fact so many causal sequences and forces at work—all of them running in parallel and each of them often affecting the course of the others—that it is hopeless to try to specify in advance what's going to happen as they jointly work themselves out. In the face of that complexity, it becomes difficult if not impossible to know with any assurance the future state of the system, except in those comparatively few cases in which the system is governed by ironclad laws of nature, such as those allowing us to predict the phases of the moon, the tides, or the position of Jupiter in tomorrow night's sky. Otherwise, forget it.

Further, it's an illusion to think that supercomputer modeling is up to the task of truly reliable crystal-ball gazing. It isn't. Witness the epidemiologists who predicted that last year's influenza season would be severe (in fact it was mild); the professional hurricane forecasters whose models told them that the last two hurricane seasons would be monsters (whereas instead they were wimps). Certain systems in nature, it seems, are computationally irreducible phenomena, meaning that there is no way of knowing the outcome short of waiting for it to happen.

Formerly, when I heard or read a prediction, I believed it. Nowadays I just roll my eyes, shake my head, and turn the page.

NICK BOSTROM

Philosopher, University of Oxford; author of Anthropic Bias: Observation Selection Effects in Science and Philosophy

Everything

For me, belief is not an all-or-nothing thing—believe or disbelieve, accept or reject. Instead, I have degrees of belief, a subjective probability distribution over different possible ways the world could be. This means I am constantly changing my mind about all sorts of things, as I reflect or gain more evidence. While I don't always think explicitly in terms of probabilities, I often do so when I give careful consideration to some matter. And when I reflect on my own cognitive processes, I must acknowledge the graduated nature of my beliefs.

The commonest way in which I change my mind is by concentrating my credence function on a narrower set of possibilities than before. This occurs every time I learn a new piece of information. Since I started my life knowing virtually nothing, I have changed my mind about virtually everything. For example, not knowing a friend's birthday, I assign a 1/365 chance (approximately) of it being August 11. After she tells me that August 11 is her birthday, I assign that date a probability of close to 100 percent. (Never exactly 100 percent, for there is always a nonzero probability of miscommunication, deception, or other error.)

It can also happen that I change my mind by smearing out my credence function over a wider set of possibilities. I might forget the exact date of my friend's birthday but remember that it is sometime in the summer. The forgetting changes my credence function from being almost entirely concentrated on August 11 to being spread out more or less evenly over the summer months. After this change of mind, I might assign a 1 percent probability to my friend's birthday being August 11.

My credence function can become more smeared out not only by forgetting but also by learning—learning that what I previously took to be strong evidence for some hypothesis is in fact weak or misleading evidence. (This type of belief change can often be mathematically modeled as a narrowing rather than a broadening of credence function, but the technicalities of this are not relevant here.) For example, over the years I have become moderately more uncertain about the benefits of medicine, nutritional supplements, and much conventional health wisdom. This belief change has come about as a result of several factors. One of the factors is that I have read some papers that cast doubt on the reliability of the standard methodological protocols used in medical studies and their reporting. Another factor is my own experience of following up on MEDLINE some of the exciting medical findings reported in the media—almost always, the search of the source literature reveals a much more complicated picture, with many studies showing a positive effect, many showing a negative effect, and many showing no effect. A third factor is the arguments of a health economist friend of mine, who takes a dim view of the marginal benefits of medical care.

Typically, my beliefs about big issues change in small steps. Ideally, these steps should approximate a random walk, like the stock market. It should be impossible for me to predict how my beliefs on some topic will change in the future. If I believed that a year hence I would assign a higher probability to some hypothesis than I do today—why, in that case, I could raise the probability right away. Given knowledge of what I will believe in the future, I would defer to the beliefs of my future self, provided I think my future self will be better informed than I am now and at least as rational.

I have no crystal ball to show me what my future self will believe. But I do have access to many other selves who are better informed than I am on many topics. I can defer to experts. Provided they are unbiased and are giving me their honest opinion, I should perhaps always defer to people who have more information than I do—or to some weighted average of expert opinion if there is no consensus. Of course, the proviso is a big one: Often I have reason to disbelieve that other people are unbiased or that they are giving me their honest opinion. However, it is also possible that I am biased and self-deceiving. An important unresolved question is how much epistemic weight a wannabe Bayesian thinker should give to the opinions of others. I'm looking forward to changing my mind on that issue, hopefully by my credence function becoming concentrated on the correct answer.

GINO SEGRE

Physicist, University of Pennsylvania; author of Faust in Copenhagen: A Struggle for the Soul of Physics

The Universe's Escape Velocity

The first topic you treat in freshman physics is showing how a ball shot straight up out of the mouth of a cannon will reach a maximum height and then fall back to Earth, unless its initial velocity (known now as "escape velocity") is great enough that it breaks out of Earth's gravitational field. If that happens, its final velocity is, however, always less than its initial one. Calculating escape velocity may not be relevant for cannonballs, but it certainly is for rocket ships.

The situation with the explosion we call the Big Bang is obviously more complicated, but really not that different—or so I thought. The standard picture said that there was an initial explosion, space began to expand, and galaxies moved away from one another. The density of matter in the universe determined whether the Big Bang would eventually be followed by a Big Crunch or the celestial objects would continue to move away from one another with decreasing acceleration. In other words, one could calculate the universe's escape velocity. Admittedly the discovery of dark matter, an unknown quantity far more

abundant than known matter, seriously altered the framework—
but not in a fundamental way, since dark matter was, after all,
still matter even if its identity was unknown.

This picture changed in 1998, with the announcement by
two teams, working independently, that the rate of acceleration of the universe's expansion was increasing, not decreasing.
It was as if freshman physics' cannonball miraculously moved
faster and faster as it left Earth. There was no possibility of a Big
Crunch, in which the universe would collapse back on itself.
The groups' analyses, based on observing distant stars of known
luminosity, type 1a supernovae, was solid. *Science* magazine
dubbed it 1998's Discovery of the Year.

The cause of this apparent gravitational repulsion is not
known. Called dark energy, to distinguish it from dark matter,
it appears to be the dominant force in the universe's expansion,
roughly three times as abundant as its dark-matter counterpart.
The prime candidate for its identity is the so-called cosmological
constant, a term introduced into the cosmic gravitation equations by Einstein to neutralize expansion but done away with by
him when astronomer Edwin Hubble reported that the universe
was in fact expanding.

Finding a theory that will successfully calculate the magnitude of this cosmological constant—assuming this is indeed the
cause of the accelerating expansion—is perhaps the outstanding
problem in the conjoined areas of cosmology and elementary particle physics. Despite many attempts, success does not seem to be
in sight. If the cosmological constant is not the answer, an alternate explanation of the dark energy would be equally exciting.

Furthermore, the apparent present equality, to within a factor of three, of matter density and the cosmological constant
has raised a series of important questions. Since matter density
decreases rapidly as the universe expands (matter per volume
decreases as volume increases) and the cosmological constant

does not, we seem to be living in that privileged moment of the universe's history when the two factors are roughly equal. Is this simply an accident? Will the distant future really be one in which, with dark energy increasingly important, celestial objects have moved so far apart so quickly as to fade from sight?

The discovery of dark energy has radically changed our view of the universe. Future keenly awaited findings, such as the identities of dark matter and dark energy, will do so again.

Psychologist, University of Massachusetts, Amherst; author of
The Cognitive Brain

What We Learn in the Living Realm of Biology

I have never questioned the conventional view that a good grounding in the physical sciences is needed for a deep understanding of the biological sciences. It did not occur to me that the opposite view might also be true.

If someone were to have asked me if biological knowledge might significantly influence my understanding of our basic physical sciences, I would have denied it. Now I am convinced that the future understanding of our most important physical principles will be profoundly shaped by what we learn in the living realm of biology. What changed my mind are the relatively recent developments in the theoretical constructs and empirical findings in the sciences of the brain, the biological foundation of all thought. Progress here can cast new light on the fundamental subjective factors that constrain our scientific formulations in what we take to be an objective enterprise.

MARK PAGEL

Evolutionary Biologist, Reading University; editor of The Oxford Encyclopedia of Evolution

We Differ Genetically
More Than We Thought

The last thirty to forty years of social science have brought an overbearing censorship to the way we are allowed to think and talk about the diversity of people on Earth. People of Siberian descent, New Guinean highlanders, inhabitants of the Indian subcontinent, Caucasians, Australian aborigines, Polynesians, Africans—we are, officially, all the same. There are no races.

Flawed as the old ideas about race are, modern genomic studies reveal a surprising, compelling, and different picture of human genetic diversity. We are, on average, about 99.5 percent similar to each other genetically. This is a new figure, down from the previous estimate of 99.9 percent. To put what may seem like minuscule differences in perspective, we are somewhere around 98.5 percent similar (maybe more) to chimpanzees, our nearest evolutionary relatives.

The new figure for us, then, is significant. It derives from, among other things, many small genetic differences that emerged from studies comparing human populations. Some confer the ability in adults to digest milk, others to withstand equatorial sun; others yet confer differences in body shape or size, resistance to particular diseases, tolerance for hot or cold, how many offspring a female might eventually produce, even the production of endorphins. We also differ by surprising amounts in the numbers of copies of some of our genes.

Modern humans spread out of Africa only within the last sixty thousand to seventy thousand years—little more than the blink of an eye when stacked against the six million or so years that separate us from our great-ape ancestors. The genetic differences among us reveal a species with a propensity to form small and relatively isolated groups on which natural selection has often acted strongly to promote genetic adaptations to particular environments.

We differ genetically more than we thought, but we should have expected this: How else but through isolation can we explain a single species that speaks at least seven thousand mutually unintelligible languages around the world?

What this all means is that, like it or not, there may be many genetic differences among human populations—including differences that may even correspond to old categories of "race"— that are real differences, in the sense of making one group better than another at responding to some particular environmental problem. This in no way says one group is in general "superior" to another, or that one group should be preferred over another. But it warns us that we must be prepared to discuss genetic differences among human populations.

Professor of astrophysics, Institute for Advanced Study, Princeton University

The Limits to Analogy

I used to pride myself on the fact that I could explain almost anything to anyone, on a simple enough level, using analogies. No matter how abstract an idea in physics may be, there always seems to be some way in which we can get at least some part of the idea across. If colleagues shrugged and said, "Oh well, that idea is too complicated or too abstract to be explained in simple terms," I thought they were either lazy or not very skilled in thinking creatively around a problem. I could not imagine a form of knowledge that could not be communicated in some limited but valid approximation or other.

However, I've changed my mind, in what was for me a rather unexpected way. I still think I was right in thinking that any type of insight can be summarized to some degree, in what is clearly a correct first approximation, when judged by someone who shares in the insight. For a long time my mistake was that I had not realized how totally wrong this first approximation can come across to someone who does not share the original insight.

Quantum mechanics offers a striking example. When someone hears that there is a limit on how accurately you can simultaneously measure various properties of an object, it is tempting to think that the limitations lie in the measuring procedure and that the object itself somehow can be held to have exact values for each of those properties even if they cannot be measured. Surprisingly, that interpretation is wrong: John Bell showed that such a "hidden variables" picture is in clear disagreement with quantum mechanics. An initial attempt at explaining the measurement problem in quantum mechanics can be more misleading than not saying anything at all.

So for each insight there is at least some explanation possible, but the same explanation may then be given for radically different insights. There is nothing that cannot be explained, but there are wrong insights that can lead to explanations that are identical to the explanation for a correct but rather subtle insight.

Psychologist, Harvard University; author of Changing Minds

Wrestling with Piaget

Like many other college students, I turned to the study of psychology for personal reasons. I wanted to understand myself better, and that understanding focused on issues of personality and motivation. And so I read the works of Freud, and I was privileged to have as my undergraduate tutor the psychoanalyst Erik Erikson, himself a sometime pupil of Freud. But once I learned about new trends in psychology, through contacts with another mentor, Jerome Bruner, I turned my attention to the operation of the mind in a cognitive sense—and I've remained at that post ever since.

The giant at the time, the middle 1960s, was Jean Piaget. Though I met and interviewed him a few times, Piaget really functioned for me as a paragon, someone who serves as a virtual teacher and point of reference. Piaget, I thought, had identified the most important question in cognitive psychology (How does the mind develop?); displayed brilliant methods of observation and experimentation; and put forth a convincing picture of development—a set of general cognitive operations that unfold in the course of essentially lockstep, universally occurring stages.

I wrote my first books about Piaget; saw myself as carrying on the Piagetian tradition in my own studies of artistic and symbolic development—two areas he had not focused on; and even defended him vigorously in print against those who dared to critique his approach and claims.

Yet now, forty years later, I have come to realize that the bulk of my scholarly career has been a critique of the principal claims that Piaget put forth. As to the specifics of how I changed my mind:

Piaget believed in general stages of development that cut across contents (e.g., space, time, number). I now believe that each area of content has its own rules and operations and I am dubious about the existence of general stages and structures.

Piaget believed that intelligence was a single general capacity that developed pretty much in the same way across individuals. I now believe that humans possess a number of relatively independent intelligences and these can function and interact in idiosyncratic ways.

Piaget was not interested in individual differences; he studied the "epistemic subject." Most of my work has focused on individual differences, with particular attention to those with special talents or deficits, and unusual profiles of abilities and disabilities.

Piaget assumed that the newborn had a few basic biological capacities—like sucking and looking—and two major processes of acquiring knowledge, which he called assimilation and accommodation. Nowadays, with many others, I assume that human beings possess considerable innate or easily elicited cognitive capacities and that Piaget greatly underestimated the power of this inborn cognitive architecture.

Piaget downplayed the importance of historical and cultural factors: Cognitive development consisted of the growing child experimenting, largely on her own, with the physical (and, minimally, the social) world. I see development as permeated from the first by contingent forces pervading the time and place of origin.

Finally, Piaget saw language and other symbols systems (graphic, musical, bodily, and so on) as manifestations—almost epiphenomena—of a single cognitive motor. I see each of these systems as having its own origins and being heavily colored by the particular uses to which a systems is put in one's own culture and one's own time.

Why I changed my mind is an issue principally of biography. Some of the change has to do with my own choices (I worked for twenty years with brain-damaged patients) and some with the Zeitgeist—I was strongly influenced by the ideas of Noam Chomsky and Jerry Fodor and by empirical discoveries in psychology and biology.

Still, I consider Piaget to be the giant of the field. He raised the right questions, he developed exquisite methods, and his observations of phenomena have turned out to be robust. It's a tribute to Piaget that we continue to ponder these questions, even as many of us are now far more critical than we once were. Any serious scientist or scholar will change his or her mind. Put differently, we will come to agree with those with whom we used to disagree and vice versa. We differ in whether we are open or secretive about such changes of mind—and in whether we choose to attack, ignore, or continue to celebrate those with whose views we are no longer in agreement.

DONALD HOFFMAN

Cognitive scientist, University of California, Irvine; author of
Visual Intelligence

Non-Veridical Perception

I have changed my mind about the nature of perception.
I thought that a goal of perception is to estimate properties of an
objective physical world and that perception is useful precisely
to the extent that its estimates are veridical. After all, incorrect
perceptions beget incorrect actions, and incorrect actions beget
fewer offspring than correct actions. Hence, on evolutionary
grounds, veridical perceptions should proliferate.

Although the image at the eye, for instance, contains insufficient information by itself to recover the true state of the world,
natural selection has built into the visual system the correct prior
assumptions about the world and how it projects onto our retinas,
so that our visual estimates are, in general, veridical. We can verify that this is the case by deducing those prior assumptions from
psychological experiments and comparing them with the world.
Vision scientists are now succeeding in this enterprise. But we
need not wait for their final report to conclude with confidence
that perception is veridical. All we need is the obvious rhetorical
question: Of what possible use is non-veridical perception?

I now think that perception is useful because it is not veridical. The argument that evolution favors veridical perceptions is wrong, both theoretically and empirically.

It is wrong in theory because natural selection hinges on reproductive fitness, not on truth, and the two are not the same. Reproductive fitness in a particular niche might, for instance, be enhanced by reducing expenditures of time and energy in perception; true perceptions, in consequence, might be less fit than niche-specific shortcuts.

It is wrong empirically because mimicry, camouflage, mating errors, and supernormal stimuli are ubiquitous in nature, and all are predicated on non-veridical perceptions. The cockroach, we suspect, sees little of the truth but is quite fit, though easily fooled, with its niche-specific perceptual hacks. Moreover, computational simulations based on evolutionary game theory—in which virtual animals that perceive the truth compete with others that sacrifice truth for speed and energy efficiency—find that true perception generally goes extinct.

It used to be hard to imagine how perceptions could possibly be useful if they were not true. Now, thanks to technology, we have a metaphor that makes it clear: the Windows interface of the personal computer. This interface sports colorful geometric icons on a two-dimensional screen. The colors, shapes, and positions of the icons on the screen are not true depictions of what they represent inside the computer. And that is why the interface is useful. It hides the complexity of the diodes, resistors, voltages, and magnetic fields inside the computer. It allows us to effectively interact with the truth because it hides the truth.

It has not been easy for me to change my mind about the nature of perception. The culprit, I think, is natural selection. I have been shaped by it to take my perceptions seriously. After all, those of our predecessors who did not, for instance, take their tiger or viper or cliff perceptions seriously had less chance of becoming our ancestors. It is apparently a small step, though not a logical one, from taking perception seriously to taking it literally.

Unfortunately, our ancestors faced no selective pressures that would prevent them from conflating the serious with the literal: one who takes the cliff both seriously and literally avoids harm just as much as one who takes the cliff seriously but not literally. Hence our collective history of believing in the flat Earth, geocentric cosmology, and veridical perception. I should very much like to join Samuel Johnson in rejecting the claim that perception is not veridical, by kicking a stone and exclaiming, "I refute it thus." But even as my foot ached from the ill-advised kick, I would still harbor the skeptical thought: "Yes, you should have taken that rock more seriously, but should you take it literally?"

IRENE PEPPERBERG

Research associate in psychology, Harvard University; author of The Alex Studies

The Virtue of Fishing Expeditions

I've begun to rethink the way we teach students to engage in scientific research. I was trained, as a chemist, to use the classic scientific method: Devise a testable hypothesis and then an experiment to see whether the hypothesis is correct. And I was told that this method was equally valid for the social sciences. I've changed my mind that this is the best way to do science. I have three reasons:

First, and probably most important, I've learned that one often needs simply to sit and observe and learn about one's subject before even attempting to devise a testable hypothesis. What are the physical capacities of the subject? What is the social and ecological structure in which it lives? Does anecdotal evidence suggest the form the hypothesis should take? Few granting agencies are willing to provide support for this step, but it is critical to the scientific process, particularly for truly innovative research. Often, a proposal to gain observational experience is dismissed as a "fishing expedition," but how can one devise a workable hypothesis to test without first acquiring basic knowledge of the system? And how better to obtain such basic knowledge than to observe the system with no preconceived notions?

Second, I've learned that truly interesting questions often can't be reduced to a simple testable hypothesis, at least not without being somewhat absurd. "Can a parrot label objects?" may be a testable hypothesis, but it isn't very interesting. What is interesting, for example, is how that labeling compares to the behavior of a young child, exactly what type of training might enable such learning, what type of training is useless, how far such labeling can transfer across exemplars, and . . . Well, you get the picture: The exciting part is a series of interrelated questions that arise and expand almost indefinitely.

Third, I've learned that the scientific community's emphasis on hypothesis-based research leads too many scientists to devise experiments to prove, rather than test, their hypotheses. Many journal submissions lack any discussion of alternative competing hypotheses. Researchers don't seem to realize that collecting data consistent with their original hypothesis doesn't mean that it is unconditionally true. Alternatively, they buy into the fallacy that absence of evidence for something is always evidence of its absence.

I'm all for rigor in scientific research, but let's emphasize the gathering of knowledge rather than the proving of a point.

ROBERT PROVINE

Psychologist and neuroscientist, University of Maryland; *author of* Laughter: A Scientific Investigation

Gone Fishing

Mentors, paper referees, and grant reviewers have warned me on occasion about scientific fishing expeditions—the conduct of empirical research that does not test a specific hypothesis or is not guided by theory. Such "blind empiricism" was said to be unscientific, to waste time and produce useless data. Although I have never been completely convinced of the hazards of fishing, I now reject those hazards outright, with a few reservations.

I'm advocating not the collection of random facts but the use of broad-based descriptive studies to learn what to study and how to study it. Those who fish learn where the fish are, their species, number, and habits. Without the guidance of preliminary descriptive studies, hypothesis testing can be inefficient and misguided. Hypothesis testing is a powerful means of rejecting error—of trimming the dead limbs from the scientific tree—but it does not generate hypotheses or signify which are worthy of test. I'll provide two examples from my experience.

In graduate school, I became intrigued with neuroembryology and wanted to introduce it to developmental psychology, a discipline that essentially starts at birth. My dissertation was a fishing expedition that described embryonic behavior and its neurophysiological mechanism. I was exploring uncharted waters and sought advice by observing the ultimate expert, the embryo. In this and related work, I discovered that prenatal movement is the product of seizurelike discharges in the spinal cord, not the brain; that the spinal discharges occurred spontaneously, not in response to sensory stimuli; that the function of movement was to sculpt joints, not to shape postnatal behavior, such as walking, and to regulate the number of motor neurons. Remarkable!

But decades later, this and similar work is largely unknown to developmental psychologists, who have no category for it. The traditional psychological specialties of perception, learning, memory, motivation, and the like are not relevant during most of the prenatal period. The finding that embryos are profoundly unpsychological beings, guided by unique developmental priorities and processes, is not appreciated by theory-driven developmental psychologists. When the fishing expedition indicates that there is no appropriate spot in the scientific filing cabinet, it may be time to add another drawer.

Years later and unrepentant, I embarked on a new fishing expedition, this time in pursuit of the human universal of laughter—what it is, when we do it, and what it means. In the spirit of my embryonic research, I wanted the expert to define my agenda—a laughing person. Explorations of research-funding with administrators at a federal agency were unpromising. One linguist patiently explained that my project "had no obvious implications for any of the major theoretical issues in linguistics." Another, a speech scientist, noted that "laughter isn't speech, and therefore has no relevance to my agency's mission."

Ultimately, my atheoretical and largely descriptive work provided many surprises and counterintuitive findings. For example, laughter, like crying, is not consciously controlled, contrary to literature suggesting that we speak "ha-ha" as we would choose a word in speech. Most laughter is not a response to humor. Laughter and speech are controlled by different brain mechanisms, with speech dominating laughter. Contagious laughter is the product of neurologically programmed social behavior. Contrasts between chimpanzee and human laughter reveal why chimpanzees can't talk (inadequate breath control) and the evolutionary event necessary for the selection for human speech (bipedality).

Whether I was exploring embryonic behavior or laughter, fishing expeditions guided me down the appropriate empirical path, provided unanticipated insights, and prevented flights of theoretical fancy. Contrary to lifelong advice, when planning a new research project, I always start by going fishing.

CHARLES SEIFE

Professor of journalism, New York University; former writer for Science; *author of* Decoding the Universe

Science and Democracy

I used to think that a modern, democratic society had to be a scientific society.

After all, the scientific revolution and the American Revolution were forged in the same flames of the Enlightenment. Naturally, I thought, a society that embraces the freedom of thought and expression of a democracy would also embrace science.

However, when I first started reporting on science, I quickly realized that science didn't spring up naturally in the fertile soil of the young American democracy. Americans were extraordinary innovators—wonderful tinkerers and engineers—but you can count the great nineteenth-century American physicists on one hand and have two fingers left over. The United States owes its scientific tradition to aristocratic Europe's universities (and to its refugees), not to any native drive.

In fact, science clashes with the democratic ideal. Though it is meritocratic, it is practiced in the elite and effete world of academe, leaving the vast majority of citizens unable to contribute to it in any meaningful way. Science is about freedom of thought, yet at the same time it imposes a tyranny of ideas.

In a democracy, ideas are protected. It's the sacred right of a citizen to hold—and to disseminate—beliefs that the majority disagrees with, ideas that are abhorrent, ideas that are wrong. However, scientists are *not* free to be completely open-minded; a scientist stops becoming a scientist if he clings to discredited notions. The basic scientific urge to falsify, to disprove, to discredit ideas clashes with the democratic drive to tolerate and protect them.

This is why even those politicians who accept evolution will never attack those politicians who don't; at least publicly, they cast evolutionary theory as a mere personal belief. Attempting to squelch creationism smacks of elitism and intolerance—it would be political suicide. Yet this is exactly what biologists are compelled to do; they exorcise falsehoods and drive them from the realm of public discourse.

We've been lucky that the transplant of science has flourished so beautifully on American soil. But I no longer take it for granted that this will continue; our democratic tendencies might get the best of us in the end.

TIMOTHY TAYLOR

Archaeologist, University of Bradford, U.K.; author of The Buried Soul: How Humans Invented Death

The Trouble with Relativism

Where once I would have striven to see Incan child sacrifice "in their terms," I am increasingly committed to seeing it in ours. Where once I would have directed attention to understanding a past cosmology of equal validity to my own, I now feel the urgency to go beyond a culturally attuned explanation and reveal cold sadism, deployed as a means of social control by a burgeoning imperial power.

In Cambridge at the end of the 1970s, I began to be inculcated with the idea that understanding the internal logic and value system of a past culture was the best way to do archaeology and anthropology. The challenge was to achieve this through sensitivity to context, classification, and symbolism. A pot was no longer just a pot but a polyvalent signifier, with a range of case-sensitive meanings. A rubbish pit was no longer an unproblematic heap of trash but a semiotic entity embodying concepts of contagion and purity, sacred and profane. A ritual killing was not to be judged bad but considered valid within a different worldview.

Using such "contextual" thinking, I no longer saw a lump of slag found in a 5000 BC female grave in Serbia as chance-contaminant, by-product garbage from making copper jewelry. Rather, it was a kind of poetic statement bearing on the relationship between biological and cultural reproduction. Just as births in the Vinča culture were attended by midwives who also delivered the warm but useless slab of afterbirth, so Vinča-culture ore was heated in a clay furnace that gave birth to metal. From the furnace—known from many ethnographies to have projecting clay breasts and a graphically vulvic stoking opening—the smelters delivered technology's baby. With it came a warm but useless lump of slag. Thus the slag in a Vinča woman's grave, far from being accidental trash, hinted at a complex symbolism of gender, death, and rebirth.

So far, so good. Relativism worked as a way toward understanding that our industrial waste was not theirs and their idea of how a woman should be appropriately buried not ours. But what happens when relativism says that our concepts of right and wrong, good and evil, kindness and cruelty, are inherently inapplicable? Relativism self-consciously divests itself of a series of anthropocentric and anachronistic skins—modern, white, Western, male-focused, individualist, scientific (or "scientistic")—to say that the recognition of such value-concepts is radically unstable, the "objective" outsider opinion a worthless myth.

My colleague Andy Wilson and our team have recently examined the hair of sacrificed children found on some of the high peaks of the Andes. Contrary to historical chronicles claiming that being ritually killed to join the mountain gods was an honor that the Incan rulers accorded only to their own privileged offspring, diachronic isotopic analyses along the scalp hairs of victims indicate they belonged to peasant children, who, a year before death, were given the outward trappings of high status and a much-improved diet in order to make them acceptable offerings. Thus we see past the self-serving accounts of those of the indigenous elite who survived on into Spanish rule. We now understand that the central command in Cuzco engineered the high-visibility sacrifice of children drawn from newly subject populations. And we can guess that this was a means to social control during the massive shock-and-awe-style imperial expansion southward into what became Argentina.

But the relativists demur from this understanding and have painted us as culturally insensitive, ignorant scientists (the last label a clear pejorative). For them, our isotope work is informative only as it reveals "the inner fantasy life of, mostly, Euro-American archaeologists, who can't possibly access the inner cognitive/cultural life of those Others." The capital "O" is significant. Here we have what the journalist Julie Burchill mordantly unpacked as "the ever-estimable Other"—the albatross that post-Enlightenment and, more important, postcolonial scholarship must wear round its neck as a sign of penance.

We need relativism as an aid to understanding past cultural logic, but it does not free us from a duty to discriminate morally and to understand that there are regularities in the negatives of human behavior as well as in its positives. In this case, it seeks to ignore what Victor Nell has described as "the historical and cross-cultural stability of the uses of cruelty for punishment, amusement, and social control." By denying the basis for a consistent underlying algebra of positive and negative, yet consistently claiming the necessary rightness of the internal cultural conduct of "the Other," relativism steps away from logic into incoherence.

LEON LEDERMAN

Physicist and Nobel laureate; director emeritus of Fermilab; coauthor (with Christopher T. Hill) of Symmetry and the Beautiful Universe

Political Science

My academic experience, mainly at Columbia University from 1946 to 1978, instilled the following firm beliefs:

The role of the professor, reflecting the mission of the university, is research and dissemination of the knowledge gained; however, the professor has many citizenship obligations—to his community, state, and nation; to his university; to his field of research, e.g., physics; to his students. In this last case, one must add "to the content knowledge transferred, the moral and ethical concerns that science brings to society."

So scientists have an obligation to communicate their knowledge, to popularize and, whenever relevant, bring their knowledge to bear on the issues of the time. Additionally, scientists play a large role in advisory boards and systems, from the President's Science Advisory Committee all the way to local school boards and PTAs. I have always believed that the above menu more or less covered all the obligations and responsibilities of the scientist but that the scientist's most sacred obligation is to continue to do science. Now I know that I was dead wrong.

Taking even a cursory stock of current events, I am driven to the ultimately wise advice of my Columbia mentor, I. I. Rabi, who, in his many corridor bull sessions, urged his students to run for public office and get elected. He insisted that to be an advisor (he was an advisor to Oppenheimer at Los Alamos, later to Eisenhower and to the Atomic Energy Commission) was ultimately an exercise in futility and that the power belonged to those who are elected. Then, we thought the old man was bonkers. But today . . .

Just look at our national and international dilemmas: global climate change (U.S. booed in Bali); nuclear weapons (seventeen years after the end of the Cold War, the U.S. has over seven thousand nuclear weapons, many poised for instant flight); stem-cell research (still hobbled by White House obstacles). Our basic research and science education are rated several nations below those of Lower Slobovia; our national deficit will burden the nation for generations; a wave of religious fundamentalism, an endless war in Iraq, and the growing security restrictions on our privacy and freedom (excused by an even more endless and mindless war on terrorism) seem to be paralyzing the Congress. We need to elect people who can think critically.

A Congress overwhelmingly dominated by lawyers and MBAs makes no sense in this twenty-first century, in which almost all issues have a science and technology aspect. We need a national movement to seek out scientists and engineers who have demonstrated the required management and communication skills. And we need a strong consensus of mentors who would insist that the need for wisdom and knowledge in the Congress must have a huge priority.

DAN SPERBER

Social and cognitive scientist; Directeur de Recherche, Centre National de la Recherche Scientifique, France; author of Explaining Culture: A Naturalistic Approach

The Paleolithic Mind

As a student, I was influenced by Claude Lévi-Strauss and even more by Noam Chomsky. Both of them dared talk about "human nature," when the received view was that there was no such thing. In my own work, I argued for a naturalistic approach in the social sciences. I took for granted that human cognitive dispositions were shaped by biological evolution and more specifically by Darwinian selection. While I did occasionally toy with evolutionary speculations, I failed to see at the time how they could play more than a marginal role in the study of human psychology and culture.

Luckily, in 1987, I was asked by Jacques Mehler, the founder and editor of *Cognition*, to review a very long article intriguingly entitled "The Logic of Social Exchange: Has Natural Selection Shaped How Humans Reason?" In most experimental psychology articles, the theoretical sections are short and relatively shallow. Here, on the other hand, the young author, Leda Cosmides, was arguing in an altogether novel way for an ambitious

theoretical claim: The forms of cooperation unique to and characteristic of humans could have evolved, she maintained, only
if there had also been, at a psychological level, the evolution of
a mental mechanism tailored to understand and manage social
exchanges and in particular to detect cheaters. Moreover, this
mechanism could be investigated by means of standard reasoning experiments.

This is not the place to go into the details of the theoretical argument—which I found and still find remarkably insight-
ful—or of the experimental evidence, which I have criticized
in detail with experiments of my own as inadequate. Whatever
its shortcomings, this was an extraordinarily stimulating paper,
and I strongly recommended acceptance of a revised version.
The article was published in 1989 and the controversies it stirred
have not yet abated.

Reading the work of Leda Cosmides and of John Tooby, her
collaborator (and husband), meeting them shortly afterward,
and initiating a conversation with them that has never ceased
made me change my mind. I had known that we could reflect
on the mental capacities of our ancestors on the basis of what
we know of our minds; I now understood that we can also draw
fundamental insights about our present minds through reflecting on the environmental problems and opportunities that have
exerted selective pressure on our Paleolithic ancestors.

Ever since, I have tried to contribute to the development
of evolutionary psychology, to the surprise and dismay of some
of my more standard-social-science friends and also of some
evolutionary psychologists, who see me more as a heretic than
a genuine convert. True, I have no taste or talent for orthodoxy;
moreover, I find much of the work done so far under the label
"evolutionary psychology" rather disappointing. Evolutionary
psychology will succeed to the extent that it causes cognitive psychologists to rethink central aspects of human cognition in an

evolutionary perspective—to the extent, that is, that psychology in general becomes evolutionary.

The human species is exceptional in its enormous investment in cognition and in forms of cognitive activity—language, higher-order thinking, abstraction—that are as unique to humans as echolocation is to bats. Yet more than half of all work done in evolutionary psychology today is about mate choice, a mental activity found in a great many species. There is nothing intrinsically wrong in studying mate choice, of course, and some of the work done in this area is outstanding. However, the promise of evolutionary psychology is first and foremost to help explain aspects of human psychology that are genuinely exceptional among earthly species and that in turn help explain the exceptional character of human culture and ecology. This is what has to be achieved to a much greater extent than has been the case so far, if we want other skeptical cognitive and social scientists to change their minds, too.

Philosopher, Johannes Gutenberg University Mainz, Germany; author of Being No One

There Are No Moral Facts

I have become convinced that it would be of fundamental importance to know what a good state of consciousness is. Are there forms of subjective experience which, in a strictly normative sense, are *better* than others? What states of consciousness should be illegal? What states of consciousness do we want to foster and cultivate and integrate into our societies? What states of consciousness can we force upon animals—for instance, in consciousness research? What states of consciousness do we want to show our children? And what state of consciousness do we eventually die in?

The past year has seen the rise of an important new discipline, neuroethics. This is not simply a new branch of applied ethics for neuroscience; it raises deeper issues, about selfhood, society, and the image of humankind. Neuroscience is now quickly transformed into neurotechnology. I predict that parts of neurotechnology will turn into consciousness technology. In 2002, out-of-body experiences were, for the first time, induced with an electrode in the brain of an epileptic patient. In 2007, we saw the first two studies, published in *Science*, demonstrating how the conscious self can be transposed to a location outside the physical body as experienced, noninvasively and in healthy subjects. Cognitive enhancers are on the rise. The conscious experience of will has been experimentally constructed and manipulated in a number of ways. Acute episodes of depression can be caused by direct interventions in the brain, and they have also been successfully blocked in previously treatment-resistant patients. And so on.

Whenever we understand the specific neural dynamics underlying a specific form of conscious content, we can in principle delete, amplify, or modulate this content in our minds. So shouldn't we also have a new ethics of consciousness—one that does not (only) ask what a good action is but goes directly to the heart of the matter, asking what we want to do with all this new knowledge and what the moral value of states of subjective experience is? I am a person with strong moral intuitions. I do have a lot of positive and explicit ideas about minimizing suffering, and about what interesting and valuable states of consciousness are. And I thought it should be possible to back up these intuitions with strong philosophical arguments.

Here is where I have changed my mind: There are no moral facts. Moral sentences have no truth-values. The world itself is silent; it just doesn't speak to us in normative affairs; nothing in the physical universe tells us what makes an action a good action or a specific brain-state a desirable one. Sure, we all would like to know what a good neurophenomenological configuration really is, and how we should optimize our conscious minds. But it looks like—in a more rigorous and serious sense—there is just no ethical knowledge to be had. We are alone. And if that is true, all we have to go by are the contingent moral intuitions that evolution has hard-wired into our emotional self-model. If we choose to simply go by what *feels* good, then our future is easy to predict: It will be primitive hedonism and organized religion.

MARC D. HAUSER

Psychologist and biologist, Harvard University; author of Moral Minds: How Nature Designed Our Universal Sense of Right and Wrong

Adaptation and Human Thought

Darwin is the man, and, like so many biologists, I have benefited from the prescient insights he handed us 150 years ago. The logic of adaptation has been a guiding engine of my research and my view of life. In fact, it has been difficult to view the world through any other filter. I can still recall with great vividness the day I arrived in Cambridge, in June 1992, a few months before starting my job as an assistant professor at Harvard. I was standing on a street corner, waiting for a bus to arrive, and noticed a group of pigeons on the sidewalk. There were several males displaying, head-bobbing and cooing, attempting to seduce the females. The females, however, were not paying attention. They were all turned, in Prussian-soldier formation, out toward the street, looking at the middle of the intersection, where traffic was whizzing by. There, in the intersection, was one male pigeon, displaying his heart out. Was this guy insane? Hadn't he read the handbook of natural selection. Dude, it's about survival. Get out of the street!!!

Further reflection provided the solution to this apparently mutant male pigeon. The logic of adaptation requires us to ask about the costs and benefits of behavior in an attempt to understand what the fitness payoffs might be. Even in behaviors that appear absurdly deleterious, there is often a benefit lurking. In this pigeon's case there was a benefit, and it was lurking in the females' voyeurism, their rubbernecking. The females were oriented toward this male, as opposed to the conservative guys on the sidewalk, because he was playing with danger, showing off — proving that even in the face of heavy traffic he could fly like a butterfly and sting like a bee, jabbing and jiving like the great Muhammad Ali.

The theory comes from the evolutionary biologist Amotz Zahavi, who proposed that even costly behaviors that challenge survival can evolve if they have payoffs to genetic fitness; these payoffs arrive in the currency of more matings, and ultimately more babies. Our male pigeon was showing off his handicap. He was advertising to the females that despite potential costs from Hummers and Beamers and buses, he was still walking the walk and talking the talk. The females were hooked, mesmerized by this extraordinarily macho male. Handicaps evolve because they are honest indicators of fitness, and Zahavi's theory represents the intellectual descendant of Darwin's original proposal.

I must admit, however, that in recent years I have made less use of Darwin's adaptive logic. This is not because I think the adaptive program has failed or can't continue to account for a wide variety of human and animal behavior. But with respect to questions of human and animal mind, and especially some of the unique products of the human mind (language, morality, music, mathematics), I have, well, changed my mind about the power of Darwinian reasoning.

Let me be clear about the claim here. I am not rejecting Darwin's emphasis on comparative approaches — that is, the use of phylogenetic or historical data. I still practice this approach, contrasting the abilities of humans and animals in the service of understanding what is uniquely human and what is shared. And I still think our cognitive prowess evolved and that the human brain and mind can be studied in some of the same ways we study other bits of anatomy and behavior. But where I have lost the faith, so to speak, is in the power of the adaptive program to explain or predict particular design features of human thought.

Although it is certainly reasonable to say that language, morality, and music have design features that are adaptive, that would enhance reproduction and survival, evidence for such claims is sorely missing. Further, for those who wish to argue that the evidence comes from the complexity of the behavior itself, and the absurdly low odds of constructing such complexity by chance, these arguments just don't cut it with respect to explaining or predicting the intricacies of language, morality, music, or many other domains of knowledge.

In fact, I would say that although Darwin's theory has been around and readily available for a hundred and fifty years, it has not advanced the fields of linguistics, ethics, or mathematics. This is not to say that it can't advance these fields, but whereas in the areas of economic decision making, mate choice, and social relationships the adaptive program has fundamentally transformed our understanding, the same cannot be said for linguistics, ethics, and mathematics. What has transformed these disciplines is our growing understanding of mechanism: how the mind represents the world, how physiological processes generate these representations, how the child grows these systems of knowledge.

Bidding Darwin adieu is not easy. My old friend has served me well. And perhaps one day he will again. Until then, farewell.

TODD E. FEINBERG

Professor of psychiatry and neurology, Albert Einstein College of Medicine; author of Altered Egos: How the Brain Creates the Self

The Soul

For most of my life, I viewed any notion of the "soul" as a fanciful religious invention. I agreed with the view of the late Nobel laureate Francis Crick, who, in his book *The Astonishing Hypothesis*, claimed, "A modern neurobiologist sees no need for the religious concept of a soul to explain the behavior of humans and other animals." But is the idea of a soul really so crazy and beyond the limits of scientific reason?

From the standpoint of neuroscience, it is easy to argue that Descartes is simply wrong about the separateness of brain and mind. The plain fact is that there is no scientific evidence that a self, an individual mind, or a soul could exist without a physical brain. However, there are persisting reasons why the self and the mind do not appear to be identical with, or entirely reducible to, the brain.

For example, in spite of the claims of Massachusetts physician Dr. Duncan MacDougall, who estimated through his experiments on dying humans that approximately twenty-one grams of matter—the presumed weight of the human soul—was lost upon death,* the mind cannot be objectively observed but only subjectively experienced. The subject that represents the "I" in the statement "I think, therefore I am" cannot be directly observed, weighed, or measured. And the experiences of that self—its pains and pleasures, sights and sounds—possess an objective reality only to the one who experiences them. In other words, as the philosopher John Searle puts it, the mind is "irreducibly first-person."

Yet although there are many perplexing properties of the brain, mind, and self that remain to be scientifically explained—subjectivity among them—this does not mean there must be an immaterial entity at work that explains these mysterious features. Nonetheless, I have come to believe that an individual consciousness represents an entity so personal and ontologically unique that it qualifies as something we might as well call a soul.

I am not suggesting that anything like a soul survives the death of the brain. In fact the link between the life of the brain and the life of the mind is irreducible, the one completely dependent on the other. Indeed, the danger of capturing the beauty and mystery of a personal consciousness and identity with the somewhat metaphorical designation "soul" is the tendency for that grandiose metaphor to obscure the brain's actual accomplishments. The soul is not a "thing," independent of the living brain; it is part and parcel of the brain—its most remarkable feature but nonetheless inextricably bound to its life and death.

---

* "Soul Has Weight, Physician Thinks," *New York Times*, March 11, 1907.

Mathematician; executive director of the Center for the Study of Language and Information, Stanford University; author of The Math Instinct

Mathematical Objects Exist Only Because We Do

Becoming a mathematician in the 1960s, I swallowed hook, line, and sinker the Platonist philosophy dominant at the time—that the objects of mathematics (the numbers, the geometric figures, the topological spaces, and so forth) had a form of existence in some abstract realm. Their existence was independent of our existence as living, cognitive creatures, and searching for new mathematical knowledge was a process of explorative discovery not unlike geographic exploration or sending out probes to distant planets.

I now see mathematics as something entirely different, as the creation of the (collective) human mind. As such, mathematics says as much about ourselves as it does about the external universe we inhabit. Mathematical facts are eternal truths, but they are not truths about an external universe that held before we entered the picture and will endure long after we are gone.

Rather, they are based on, and reflect, our interactions with the external environment around us.

This is not to say that mathematics is something we have freedom to invent. It's not like literature or music, where there are constraints on the form but writers and musicians exercise great creative freedom within those constraints. From the perspective of the individual human mathematician, mathematics is indeed a process of discovery. But what is being discovered is a product of the human (species)–environment interaction.

This view raises the fascinating possibility that other cognitive creatures in another part of the universe might have different mathematics. Of course, as a human, I cannot begin to imagine what that might mean. It would classify as "mathematics" only insofar as it amounted to that species' analyzing the abstract structures arising from their interactions with their environment.

This shift in philosophy has influenced the way I teach; I now stress social aspects of mathematics. But when I'm giving a specific lecture on, say, calculus or topology, my approach is entirely Platonist. We do our mathematics using a physical brain that evolved over hundreds of thousands of years by a process of natural selection to handle the physical and more recently the social environments in which our ancestors found themselves. As a result, the only way for the brain to actually do mathematics is to approach it "Platonistically," treating mathematical abstractions as physical objects that exist.

A Platonist standpoint is essential to doing mathematics, just as Cartesian dualism is virtually impossible to dispense with in doing science, or just plain communicating with one another ("one another"?). But ultimately our mathematics is just that: our mathematics, not the universe's.

DANIEL EVERETT

Researcher of Pirahã culture; chair of Languages, Literatures, and Cultures Department, professor of linguistics and anthropology, Illinois State University; author of the forthcoming Don't Sleep, There Are Snakes: Life and Language in the Amazonian Jungle

Where Does Grammar Come From?

I have occasionally irritated colleagues by remarking, "If it ain't broke, break it." At the same time, I have adhered to a value common in the day-to-day business of scientific research— namely, that changing one's mind is all right for little matters but suspect when it comes to big questions. Take a theory compatible with either conclusion *x* or conclusion *y*. First you believed *x*. Then you received new information and you believed *y*. This is a little change, and it is a natural form of learning—a change in behavior resulting from exposure to new information. But change your mind about the general theory you work with, and—at least in some fields—you are looked on as a kind of maverick, a person without proper research priorities, a pot-stirrer. Why is that? I wonder.

The stigma against major mind changes in science results from what I call homeopathic bias—the belief that scientific knowledge is built up bit by little bit, as we move cumulatively toward the truth. This bias can lead researchers to avoid concluding that their work undermines the dominant theory in any significant way. Nonhomeopathic doses of criticism can be considered not merely inappropriate but even arrogant: Somehow the researcher is superior to his or her colleagues, whose unifying conceptual scheme is now judged to be weaker than they have noticed or are willing to concede.

So any scientist publishing an article or a book about a nonhomeopathic mind-change could be committing a career-endangering act. But I love to read these kinds of books. They bother people. They bother me.

I changed my mind about this homeopathic bias. I think it is myopic for the most part. I came to this conclusion because I changed my mind regarding the largest question of my field: where language comes from. This change taught me about the empirical issues that had led to my shift and about the forces that can hold science and scientists in check if we aren't aware of them.

I believed at one time that culture and language were largely independent. Yet there is a growing body of research that suggests the opposite: Deep reflexes from culture are to be found in grammar. But if culture can exercise major effects on grammar, then the theory I had committed most of my research career to—the theory that grammar is part of the human genome and that the variations in the grammars of the world's languages are largely insignificant—was dead wrong. There did not have to be a specific genetic capacity for grammar; the biological basis of grammar could be the same as the basis of gourmet cooking, or mathematics, or medical advances: human reasoning.

Grammar had once seemed to me too complicated to derive from any general human cognitive properties. It appeared to cry out for a specialized component of the brain, or what some linguists call the language organ. But such an organ becomes implausible if we can show that it is not needed because there are other forces that can explain language as both ontogenetic and phylogenetic fact.

Many researchers have discussed the kinds of things that hunters and gatherers needed to talk about and how these influenced language evolution. Our ancestors had to talk about things and events, about relative quantities, and about the contents of the minds of their conspecifics, among other things. If you can't talk about things and what happens to them (events) or what they are like (states), you can't talk about anything. So all languages need verbs and nouns. But I have been convinced by the research of others, as well as my own, that if a language has these, then the basic skeleton of the grammar largely follows. The meanings of verbs require a certain number of nouns, and those nouns plus the verb make simple sentences, ordered in logically restricted ways. Other permutations of this foundational grammar follow from culture, contextual prominence, and modification of nouns and verbs. There are other components to grammar, but not all that many. Put like this, as I began to see things, there really doesn't seem to be much need for grammar proper to be part of the human genome, as it were. Perhaps there is even much less need for grammar as an independent entity than we might have once thought.

GARY MARCUS

Director of New York University's Infant Language Learning Center; author of Kluge: The Haphazard Construction of the Human Mind

Nothing Emerges from Scratch

When I was in graduate school, in the early 1990s, I learned two important things: that the human capacity for language was innate and that the machinery allowing human beings to learn language was special, in the sense of being separate from the rest of the human mind. Both ideas sounded great at the time, but (as far as I can tell) only one of them turns out to be true.

I still think I was right to believe in innateness, the idea that the human mind arrives, fresh from the factory, with a considerable amount of elaborate machinery. When a human embryo emerges from the womb, it has almost all the neurons it will ever have. All the basic neural structures are in place and most or all of the basic neural pathways are established. There is, to be sure, lots of learning yet to come—an infant's brain is more rough draft than final product—but anybody who still imagines the infant human mind to be little more than an empty sponge isn't in touch with the realities of modern genetics and neuroscience. Almost half our genome is dedicated to the development of brain function, and those ten thousand or fifteen thousand brain-related genes choreograph an enormous amount of biological sophistication. Noam Chomsky, whose classes I sat in on while in graduate school, was absolutely right to insist for all these years that language has its origins in the built-in structure of the mind.

But now I believe I was wrong to accept the idea that language is separate from the rest of the human mind. It's always been clear that we can talk about what we think about, but when I was in graduate school it was popular to talk about language as being acquired by a "module" or "instinct" separate from the rest of cognition, by what Chomsky called a language-acquisition device, or LAD. Its mission in life was to acquire language and nothing else.

In keeping with the idea of language as a product of a specialized inborn mechanism, we noted how quickly human toddlers acquired language and how determined they were to do so. All normal human children acquire language, not just a select few raised in privileged environments, and they manage to do so rapidly, learning most of what they need to know in the first few years of life. The average adult, by contrast, often gives up around the time he or she has to face the fourth list of irregular verbs. Combine that with the fact that some children with normal intelligence couldn't learn language and others with normal language lacked normal cognitive function, and I was convinced. Humans acquired language because they had a built-in module uniquely dedicated to that function.

Or so I thought then. By the late 1990s, I started looking beyond the walls of my own field—developmental psycholinguistics—and out toward a whole host of other fields, including genetics, neuroscience, and evolutionary biology.

The idea that most impressed me and did the most to shake me of the belief that language was separate from the rest of the mind goes back to Darwin. Not "survival of the fittest" (a phrase actually coined by Herbert Spencer) but his notion, now amply confirmed at the molecular level, that all biology is the product of what he called "descent with modification." Every species and every biological system evolves through a combination of inheritance (descent) and change (modification). Nothing—no matter how original it may appear—emerges from scratch.

Language, I ultimately realized, must be no different. It emerged quickly, in the space of a few hundred thousand years, and with comparatively little genetic change. It suddenly dawned on me that the striking fact that our genomes overlap almost 99 percent with those of chimpanzees must be telling us something: Language couldn't possibly have started from scratch. There isn't enough room in the genome, or in our evolutionary history, for it to be plausible that language is completely separate from what came before.

Instead, I have now come to believe that language must be largely a recombination of spare parts, a kind of jury-rigged kluge, built largely out of cognitive machinery that evolved for other purposes long before there was such a thing as language. If there's something special about language, it is not the parts from which it is composed but the way in which they are put together.

Neuroimaging studies seem to bear this out. Whereas we once imagined language to be produced and comprehended almost entirely by two purpose-built regions, Broca's area and Wernicke's area, we now see that many other parts of the brain are involved—for example, the cerebellum and basal ganglia—and that the classic language areas, Broca's and Wernicke's, participate in other aspects of mental life (for example, music and motor control) and have counterparts in other apes.

At the narrowest level, this means that psycholinguists and cognitive neuroscientists need to rethink their theories about what language is. But if there is a broader lesson, it is this: Although we humans in many ways differ radically from any other species, our greatest gifts are built upon a genomic bedrock we share with the many other apes that walk the Earth.

Computer scientist; researcher at the Massachusetts Institute of Technology's Center for Bits and Atoms

The Overloading of Bob

Not that long ago, I was under the impression that the basic problem of computer architecture had been solved. After all, computers got faster every year, and gradually whole new application domains emerged. There was constantly more memory available, and software hungrily consumed it. Each new computer had a bigger power supply, and more airflow to extract the increasing heat from the processor.

Now clock speeds aren't rising quite as quickly and the progress that is made doesn't seem to help our computers start up or run any faster. The traditions of the computing industry, some going as far back as the first digital computers built by John von Neumann in the 1950s, are starting to grow obsolete. The slower computers seem to get faster, and the more deeply I understand the way things actually work, the more these problems become apparent to me. They really come to light when you think about a computer as a business.

Imagine if your company or organization had one fellow (the CPU) who sat in an isolated office and refused to talk with anyone except his two most trusted deputies (the Northbridge and Southbridge), through which all the actual work the company does must be funneled. Because this one man—let's call him Bob—is so overloaded doing all the work of the entire company, he has several assistants (memory controllers) who remember everything for him. They do this through a complex system (virtual memory) of file cabinets of various sizes (physical memories), the organization over which they have strictly limited autonomy.

Because it is faster to find things in the smaller cabinets (RAM), where there is less to sift through, Bob asks them to put the most commonly used information there. But since he is constantly switching between different tasks, the assistants must swap in and out the files in the smaller cabinets with those in the larger ones whenever Bob works on something different ("thrashing"). The largest file cabinet is humongous and rotates slowly in front of a narrow slit (magnetic storage). The assistant in charge of it must simply wait for the right folder to appear in front of him before passing it along (disk latency).

Any communication with customers must be handled through a team of receptionists (I/O controllers) who don't take the initiative to relay requests to one of Bob's deputies. When Bob needs customer input to continue on a difficult problem, he drops what he is doing to chase after his deputy to chase after a receptionist to chase down the customer, thus preventing work for other customers from being done in that time.

This model is clearly horrendous, for numerous reasons. If any staff member goes out to lunch, the whole operation is likely to grind to a halt. Tasks that ought to be quite simple turn out to take a lot of time, since Bob must reacquaint himself with the issues in question. If a spy gains Bob's trust, all is lost. The only way to make the model any better without giving up and starting over is to hire people who just do their work faster and spend more hours in the office.

And yet, this is the way almost every computer in the world operates today.

It is much more sane to hire a large pool of individuals and, depending on slow-changing customer needs, organize them into business units and assign them to customer accounts. Each person keeps track of his own small workload, and everyone can work on a separate task simultaneously. If the company suddenly acquires new customers, it can recruit more staff instead of forcing Bob to work overtime. If a certain customer demands more attention than was foreseen, more people can be devoted to the effort. And perhaps most important, collaboration with other businesses becomes far more meaningful than the highly coded, formal game of telephone that Bob must play with Frank, who works in a similar position at another corporation (a server). Essentially, this is a business model problem as much as a computer science one.

These complaints only scratch the surface of the design flaws of today's computers. On an extremely low level, with voltages, charge, and transistors, energy is handled recklessly, causing tremendous heat, which would melt the parts in a matter of seconds were it not for the noisy cooling systems we find in most computers. And on a high level, software engineers have constructed a city of competing abstractions based on the fundamentally flawed CPU idea.

So I have changed my mind. I used to believe that computers were on the right track, but now I think the right thing to do is to move forward from our 1950s models to a ground-up, fundamentally distributed computing architecture. I started to use computers when I was seventeen months old and started programming them at age five, so I took the model for granted. But the current stagnation of perceptual computer performance and the counterintuitiveness of programming languages has led me to question what I was born into and wonder if there's a better way. I'm eager to help make it happen. When discontent changes your mind, that's innovation.

MAX TEGMARK

Physicist, Massachusetts Institute of Technology; researcher in precision cosmology

The Bird's-Eye View and the Frog's-Eye View

Do we need to understand consciousness to understand physics? I used to answer yes, thinking that we could never figure out the elusive Theory of Everything for our external physical reality without first understanding the distorting mental lens through which we perceive it.

After all, physical reality has turned out to be very different from the way it seems, and most of our notions about it have turned out to be illusions. The world looks as though it has three primary colors, but that number, three, tells us nothing about the world out there, merely something about our senses: that our retina has three kinds of cone cells. The world looks as though it has impenetrably solid and stationary objects, but all except a quadrillionth of the volume of a rock is empty space between particles in restless schizophrenic vibration. The world feels like a three-dimensional stage where events unfold over time, but Einstein's work suggests that change is an illusion, time being merely the fourth dimension of an unchanging spacetime that just is, never created and never destroyed, containing our cosmic history as a DVD contains a movie. The quantum world feels random, but Hugh Everett's work suggests that randomness, too, is an illusion, being simply the way our minds feel when cloned into diverging parallel universes.

The ultimate triumph of physics would be to start with a mathematical description of the world from the bird's-eye view of a mathematician studying the equations (ideally, simple enough to fit on her T-shirt) and to derive from them the "frog's-eye view" of the world, the way her mind subjectively perceives it. However, there is a third and intermediate consensus view of the world. From your subjectively perceived frog perspective, the world turns upside down when you stand on your head and disappears when you close your eyes, yet you subconsciously interpret your sensory inputs as though there is an external reality independent of your orientation, your location, and your state of mind. It is striking that although this third view involves both censorship (like rejecting dreams), interpolation (between eyeblinks), and extrapolation (like attributing existence to unseen cities) of your frog's-eye view, independent observers nonetheless appear to share this consensus view. Although the frog's-eye view looks black-and-white to a cat, iridescent to a bird seeing four primary colors, and still more different to a bee seeing polarized light, a bat using sonar, a blind person with keener touch and hearing, or the latest robotic vacuum cleaner, all can agree on whether a door is open or closed.

This reconstructed consensus view of the world that humans, cats, aliens, and future robots would all agree on is not free from some of the above-mentioned shared illusions. However, it is by definition free from illusions that are unique to biological minds, and therefore decouples from the issue of how our human consciousness works. This is why I've changed my mind: Although understanding the detailed nature of human consciousness is a fascinating challenge in its own right, it is not necessary for a fundamental theory of physics, which need "only" derive the consensus view from its equations.

In other words, what Douglas Adams called "the ultimate question of life, the universe and everything" splits cleanly into two parts that can be tackled separately: The challenge for physics is deriving the consensus view from the bird's-eye view, and the challenge for cognitive science is to derive the frog's-eye view from the consensus view. These are two great challenges for the third millennium. They are each daunting in their own right, and I'm relieved that we need not solve them simultaneously.

ROBERT SAPOLSKY

Neuroscientist, Stanford University; author of Why Zebras Don't Get Ulcers

Looking at Minds

Well, my biggest change of mind came only a few years ago. It was the outcome of a painful journey of self-discovery, where my wife and children stood behind me and made it possible, where I struggled with all my soul and all my heart and all my might. But that had to do with my realizing that Broadway musicals are not cultural travesties, so it's a little tangential here. Instead I'll focus on science.

I'm both a neurobiologist and a primatologist, and I've changed my mind about plenty of things in both of these realms. But the most fundamental change is one that transcends either of those disciplines. This was my realizing that the most interesting and important things in the life sciences are not going to be explained with sheer reductionism.

A specific change of mind concerned my work as a neurobiologist. This came about fifteen years ago, and it challenged neurobiological dogma I had learned in preschool—namely, that the adult brain does not make new neurons. This fact had always been a point of weird pride in the field: Hey, the brain

115

is so fancy and amazing that its elements are irreplaceable, not like some dumb-ass, simplistic liver that's so totally fungible it can regrow itself. What this fact also reinforced, in passing, was the dogma that the brain is set in stone very early in life—that there's all sorts of things that can't be changed once a certain time window has passed.

Starting in the 1960s, a handful of crackpot scientists cried in the wilderness about how the adult brain does make new neurons. At best, their unorthodoxy was ignored; at worst, they were punished for it. But by the 1990s, it had become clear that they were right. And adult neurogenesis has turned into the hottest subject in the field. The brain makes new neurons—makes them under interesting circumstances, fails to make them under other interesting ones. The new neurons function, are integrated into circuits, might even be required for certain types of learning. And the phenomenon is a cornerstone of a new type of neurobiological chauvinism: Part of the very complexity and magnificence of the brain is how it can rebuild itself in response to the world around it.

I'll admit that this business about new neurons was a tough one for me to assimilate. I wasn't invested enough in the whole business to be in the crowd indignantly saying, "No, this can't be true!" Instead, I just tried to ignore it: "New neurons? I can't deal with this! Turn the page." After an embarrassingly long time, enough evidence had piled up that I had to change my mind and decide I needed to deal with it after all. And it's now one of the things my lab studies.

The other change concerned my life as a primatologist, studying male baboons in East Africa. This also came in the early 1990s. I study what social behavior has to do with health, and my shtick always was that if you want to know which baboons are going to be festering with stress-related disease, look at the low-ranking ones. Rank is physiological destiny, and if you have a choice in the matter, you want to win some critical fights and become a dominant male, because you'll be healthier. My change of mind was twofold: I realized, from my own data and that of others, that being dominant has far less to do with winning fights than with social intelligence and impulse control. And I realized that while health has something to do with social rank, it has far more to do with personality and social affiliation: If you want to be a healthy baboon, don't be a socially isolated one. This particular shift has to do with the accretion of new facts, new statistical techniques for analyzing data, blah, blah. Probably most important, it has to do with the fact that I was once a hermetic twenty-two-year-old studying baboons and now, thirty years later, I've changed my mind about a lot of things in my life.

TOR NØRRETRANDERS

Science writer, consultant, and lecturer based in Copenhagen; author of The Generous Man: How Helping Others Is the Sexiest Thing You Can Do

What Is Constant in You Is Not Material

I have changed my mind about my body. I used to think of it as a kind of hardware on which my mental and behavioral software was running. Now I think of my body primarily as software.

My body is not like a typical material object, a stable thing. It is more like a flame, a river, or an eddy. Matter is flowing through it all the time. The constituents are being replaced over and over again.

A chair or a table is stable because the atoms stay where they are. The stability of a river stems from the constant flow of water through it.

Ninety-eight percent of the atoms in your body are replaced every year. Ninety-eight percent! Water molecules stay in your body for two weeks and for an even shorter time in a hot climate; the atoms in your bones stay there for a few months. Some atoms stay for years, but hardly any single atom stays with you from cradle to grave.

What is constant in you is not material. An average person takes in one and a half tons of matter every year as food, drink, and oxygen. All this matter has to learn to be you. Every year. New atoms will have to learn to remember your childhood.

These numbers have been known for fifty years or more, mostly from studies of radioactive isotopes. Physicist Richard Feynman said in 1955, "Last week's potatoes! They now can remember what was going on in your mind a year ago."

But why is this simple insight not on the all-time top-ten list of important discoveries? Perhaps because it tastes a little like spiritualism and idealism. Only the ghosts are for real? Wandering souls?

But digital media now make it possible to think of all this in a simple way. The music I danced to as a teenager has been moved from vinyl LPs to magnetic audio tapes to CDs to iPods and whatnot. The physical representation can change and is not important—as long as it is there. The music can jump from medium to medium, but it is lost if it does not have a representation. This physics of information was sorted out by Rolf Landauer in the 1960s. Likewise, our memories can move from potato-atoms to burger-atoms to banana-atoms. But the moment they are on their own, they are lost.

We reincarnate ourselves all the time. We constantly give our personality new flesh. I keep my mental life alive by making it jump from atom to atom. A constant flow. Never the same atoms, always the same river. No flow, no river. No flow, no me.

This is what I call permanent reincarnation—software replacing its hardware all the time. Atoms replacing atoms all the time. Life. This is very different from religious reincarnation, with souls jumping from body to body or souls sitting out there, waiting for a body to go home to.

There has to be material continuity for permanent reincarnation to be possible. The software is what is preserved, but it cannot live on its own. It has to jump from molecule to molecule, always incarnating.

I have changed my mind about the stability of my body: It keeps changing all the time. Or I could not stay the same.

HELEN FISHER

Research professor, Department of Anthropology, Rutgers University; author of Why We Love

The Four-Year Itch

When asked why all of her marriages failed, anthropologist Margaret Mead is said to have replied, "I beg your pardon. I have had three marriages and none of them was a failure."

There are many people like Mead. Some 90 percent of Americans marry by middle age. When I looked at United Nations data on ninety-seven other societies, I found that more than 90 percent of men and women eventually wed in the vast majority of these cultures, too. Moreover, most human beings around the world are monogamists. Yet, almost everywhere, people have devised social or legal means to untie the knot, and where they can divorce, and remarry, many do.

I had long suspected that this human habit of serial monogamy had evolved for some biological purpose. Planned obsolescence of the pairbond? Perhaps the mythological "seven-year itch" evolved millions of years ago to enable a bonded pair to rear two children through infancy together. If each departed after about seven years to seek "fresh features," as Lord Byron put it, both would have ostensibly reproduced themselves and both could breed again, creating more genetic variety in their young.

So I began to cull divorce data on fifty-eight societies collected since 1947 by the Statistical Office of the United Nations. My mission: to prove that the seven-year itch was a worldwide biological phenomenon associated in some way with rearing young.

Not to be. My intellectual transformation came while I was examining these divorce statistics in a rambling cottage—a shack, really—on the Massachusetts coast one August morning. I regularly got up around 5:30, went to a tiny desk that overlooked the deep woods, and pored over the pages I had xeroxed from the U.N. Demographic Yearbooks. But in country after country, decade after decade, divorces tended to peak (the divorce mode) during and around the fourth year of marriage. There were variations, of course: Americans tended to divorce between the second and third year of marriage, for example. Interestingly, this corresponds with the normal duration of intense early-stage, romantic love—often about eighteen months to three years. Indeed, in a 2007 Harris poll, 47 percent of American respondents said they would depart an unhappy marriage when the romance wore off unless they had conceived a child.

Nevertheless there was no denying it: Among these hundreds of millions of people from vastly different cultures, three patterns kept emerging. Divorces regularly peaked during and around the fourth year after wedding. Divorces peaked among couples in their late twenties. And the more children a couple had, the less likely they were to divorce: Some 39 percent of worldwide divorces occurred among couples with no dependent children; 26 percent occurred among those with one child; 19 percent occurred among couples with two children; and 7 percent occurred among couples with three young.

I was so disappointed. I mulled about this endlessly. Why do so many men and women divorce during and around the four-year mark, at the height of their reproductive years, and often with only a single child? It seemed like such an unstable reproductive strategy. Then suddenly I had an *ah-ha!* moment: Women in hunting and gathering societies breastfeed around the clock, eat a low-fat diet, and get a lot of exercise—habits that tend to inhibit ovulation. As a result, they regularly space their children about four years apart. Thus, the modern duration of many marriages—about four years—conforms to the traditional period of human birth spacing: four years.

Perhaps human parental bonds originally evolved to last only long enough to raise a single child through infancy, about four years, unless a second infant was conceived. By age five, a youngster could be reared by its mother and a host of relatives. Equally important, both parents could choose a new partner and bear more varied young.

My new theory fit nicely with data on other species. Only about 3 percent of mammals form a pairbond to rear their young. Take foxes. The vixen's milk is low in fat and protein; she must feed her kits constantly, and she will starve unless the dog fox brings her food. So foxes pair in February and rear their young together. But when the kits leave the den in midsummer, the pairbond breaks up. Among foxes, the partnership lasts only through the breeding season. This pattern is common in birds. Among the more than eight thousand avian species, some 90 percent form a pairbond to rear their young, but most do not pair for life. A male and female robin, for example, form a bond in the early spring and rear one or more broods together. But when the last of the fledgling fly away, the pairbond breaks up.

Like pairbonding in many other creatures, humans have probably inherited a tendency to love and love again in order to create more genetic variety in the young. We aren't puppets on a string of DNA, of course. Today some 57 percent of American marriages last for life. But deep in the human spirit is a restlessness in long relationships, born of a time long gone, when, as poet John Dryden put it, "wild in wood the noble savage ran."

Science writer; contributing editor of Astronomy *magazine*

Shifting from Forward into Reverse

When I was twenty-one, I began working for the Union of Concerned Scientists in Cambridge, Massachusetts. I was still an undergraduate at the time, planning on doing a brief research stint in energy policy before finishing college and heading to graduate school in physics. That "brief research stint" lasted seven years, off and on, and I never did make it to graduate school. But the experience was instructive.

When I started at UCS in the 1970s, nuclear power safety was a hot topic, and I squared off in many debates against nuclear proponents from utility companies, nuclear engineering departments, and so forth regarding reactor safety, radioactive wastes, and the viability of renewable energy alternatives. The next issue I took on for UCS was the nuclear arms race, which was equally polarized. (The neocons of that day weren't neo, just cons.) As with nuclear safety, there was essentially no common ground between the two sides. Each faction was trying to do the other in, through oral rhetoric and tendentious prose, always looking for new material to buttress their case or undermine that of their opponents.

Even though the organization I worked for was called the Union of Concerned Scientists, and even though many of the staff members there referred to me as a scientist (despite my lack of academic credentials), I knew that what I was doing was not science. Nor were the many physics PhDs in arms control and energy policy doing science. In the back of my mind, I assumed that "real science" was different—that scientists are guided by facts rather than by ideological positions, personal rivalries, and the like.

In the decades since, I've learned that while this may be true in many instances, often it's not. When it comes to the biggest, most contentious issues in physics and cosmology—such as the validity of inflationary theory, string theory, or the multiverse/landscape scenario—the image of the objective truth-seeker standing above the fray, calmly sifting through the evidence without preconceptions or prejudice, may be less accurate than the adversarial model of, say, our justice system. Both sides, to the extent that there are sides on these matters, are constantly assembling their briefs, trying to convince themselves as well as the jury while looking for flaws in the arguments of opposing counsel.

This fractionalization may stem from scientific intuition, political or philosophical differences, personal grudges, or pure academic competition. It's not surprising that this happens, nor is it necessarily a bad thing. In fact, it's my impression that this approach works pretty well in the law and in science both. It means that, on the big things at least, the science will be vetted; it has to withstand scrutiny, pass muster.

But it's not a cold, passionless exercise: At its heart, science is a human endeavor, carried out by people. When the questions are truly ambitious, it takes a great personal commitment to make any headway—a big investment in energy and in emotion as well. I know from having met with many of the leading researchers that the debates can get heated, sometimes uncomfortably so. More important, when you're engaged in an epic struggle—trying to put together a theory of broad sweep—it may be difficult, if not impossible, to keep an open mind; you may be well beyond that stage, having long since cast your lot with a particular line of reasoning. After making an investment over the course of many years, it's natural to want to protect it. That doesn't mean you can't change your mind—I know of several cases where this has occurred—but it's never easy to shift from forward into reverse.

Although I haven't worked as a scientist in any of these areas, I have written about many of the "big questions," and I know how hard it is to get all the facts lined up so they fit together in something resembling an organic whole. Doing that, even as a mere scribe, involves single-minded exertion, and in the process, the issues almost take on a life of their own—at least while you're actively thinking about them. Before long, you've moved onto the next story, and the excitement of the former recedes. As the urgency fades, you start wondering why you felt so strongly about "the landscape" or "eternal inflation" or whatever it was that took over your desk some months ago.

It's different, of course, for the researchers, who may stake out an entire career—or big chunks thereof—in a certain field. They're obliged to keep abreast of all that's going on of note, which means their interest is continually renewed. As new data comes in, they try to see how it fits in with the pieces of the puzzle they're already grappling with. Or if something significant emerges from the opposing camp, they may instinctively seek out the weak spots, trying to see how those guys messed up this time.

It's possible, of course, that the day may come when, try as you might, you cannot find the weak spots in the other guy's story. After many attempts and an equal number of setbacks, you may ultimately have to accede to the view of an intellectual, if not a personal, rival. Not that you want to, but rather because you can't see any way around it. You may chalk it up as a defeat, something that will, you hope, build character down the road. But in the grand scheme of things it's more of a victory—a sign that sometimes our adversarial system of science actually works.

Physicist, Albert Einstein Professor of Science, Princeton University; coauthor (with Neil Turok) of Endless Universe: Beyond the Big Bang

A Universe That Is Random
and Unpredictable

What created the structure of the universe?

Most cosmologists would say the answer is "inflation," and until recently I would have been among them. But facts have changed my mind—and I now feel compelled to seek a new explanation that may or may not incorporate inflation.

The idea always seemed incredibly simple. Inflation is a period of rapid accelerated expansion that can transform the chaos emerging from the Big Bang into the smooth, flat homogeny observed by astronomers. If one likens the conditions following the Bang to a wrinkled and twisted sheet of perfectly elastic rubber, then inflation corresponds to stretching the sheet at faster-than-light speeds until no vestige of its initial state remains. The "inflationary energy" driving the accelerated expansion then decays into the matter and radiation seen today, and the stretching slows to a modest pace that allows the matter to condense into atoms, molecules, dust, planets, stars, and galaxies.

I would describe this version as the classical view of inflation, in two senses. First, this is the historical picture of inflation as initially introduced and now appearing in most popular descriptions. Second, this picture is founded on the laws of classical physics, assuming quantum physics plays a minor role. Unfortunately, this classical view is dead wrong. Quantum physics turns out to play an absolutely dominant role in shaping the inflationary universe. In fact, inflation amplifies the randomness inherent in quantum physics to produce a universe that is random and unpredictable.

This realization has come slowly. Ironically, the role of quantum physics was believed to be a boon to the inflationary paradigm when it was first considered twenty-five years ago by several theorists, including me. The classical picture of inflation could not be strictly true, we recognized, or else the universe would be so smooth after inflation that galaxies and other large-scale structures would never form. However, inflation ends through the quantum decay of inflationary energy into matter and radiation. The quantum decay is analogous to the decay of radioactive uranium, in which there is some mean rate of decay but inherent unpredictability as to when any particular uranium nucleus will decay. Long after most uranium nuclei have decayed, there remain some nuclei that have yet to fission.

Similarly, inflationary energy decays at slightly different times in different places, leading to spatial variations in the temperature and matter density after inflation ends. The "average" statistical pattern appears to agree beautifully with the pattern of microwave background radiation emanating from the earliest stages of the universe and to produce just the pattern of nonuniformities needed to explain the evolution and distribution of galaxies. The agreement between theoretical calculation and observations is a celebrated triumph of the inflationary picture.

But is this really a triumph? Only if the classical view is correct. In the quantum view, it makes no sense to talk about an "average" pattern. The problem is that, as in the case of uranium nuclei, there always remain some regions of space in which the inflationary energy has not yet decayed into matter and radiation. Although one might guess that the undecayed regions are rare, they expand so much faster than those that have decayed that they soon overtake the volume of the universe. The patches where inflationary energy has decayed and galaxies and stars have evolved become the oddity—rare pockets surrounded by space that continues to inflate away.

The process repeats itself over and over, with the number of pockets and the volume of surrounding space increasing from moment to moment. Due to random quantum fluctuations, pockets with all kinds of properties are produced—some flat, some curved; some with variations in temperature and density like what we observe, but some not; some with forces and physical laws like those we experience, but some with different laws. The alarming result is that there are an infinite number of pockets of each type, and despite over a decade of attempts to avoid the situation, no mathematical way of deciding which is more probable has been shown to exist.

Curiously, this unpredictable "quantum view" of inflation has not yet found its way into the consciousness of many astronomers working in the field, let alone the greater scientific community or the public at large.

One often reads that recent measurements of the cosmic microwave background or the large-scale structure of the universe have verified a prediction of inflation. This invariably refers to a prediction based on the naive, classical view. But if the measurements ever come out differently, this could not rule out inflation. According to the quantum view, there are invariably pockets with matching properties.

And what of the theorists who have been developing the inflationary theory for the last twenty-five years? Some, like me, have been in denial, harboring the hope that a way can be found to tame the quantum effects and restore the classical view. Others have embraced the idea that cosmology may be inherently unpredictable, although this group is also vociferous in pointing out how observations agree with the (classical) predictions of inflation.

It may have taken me longer to accept inflation's quantum nature than it should have, but, now that facts have changed my mind, I cannot go back again. Inflation does not explain the structure of the universe. Perhaps some enhancement can explain why the classical view works so well, but then it will be the enhancement, rather than inflation itself, that explains the structure of the universe. Or maybe the answer lies beyond the Big Bang. Some of us are considering the possibility that the evolution of the universe is cyclic and that the structure was set by events that occurred before the Big Bang. One of the attractive features of the cyclic picture is that quantum physics never dominates and there is no production of pocket universes with different properties. Instead, quantum physics is kept under control, acting as a small perturbation of the classical view. As a result, the universe is kept uniform and flat and has nearly the same properties everywhere.

RODNEY BROOKS

Panasonic Professor of Robotics, Massachusetts Institute of Technology; chief technical officer of iRobot Corporation; author of Flesh and Machines: How Robots Will Change Us

We Need New Metaphors

Our science, including mine, treats living systems as mechanisms at multiple levels of abstraction. As we talk about how one biomolecule docks with another, our explanations are purely mechanistic, and our science never invokes "and then the soul intercedes and gets them to link up." The underlying assumption of molecular biologists is that their level of mechanistic explanation is ultimately adequate for high-level mechanistic descriptions such as physiology and neuroscience to build on as a foundation.

Those of us who are computer scientists by training—and, I'm afraid, many collaterally damaged scientists of other stripes— tend to use computation as the mechanistic level of explanation for how living systems behave and "think." I originally gleefully embraced the computational metaphor.

If we look back over recent centuries, we see the brain described as a hydrodynamic machine, clockwork, or steam engine. When I was a child, in the 1950s, I read that the human brain was a telephone switching network. Later it became a digital computer and then a massively parallel digital computer. A few years ago, someone put up a hand after a talk I gave at the University of Utah and asked a question I had been waiting for for a couple of years: "Isn't the human brain just like the World Wide Web?" The brain always seems to be one of the most advanced technologies we humans currently have.

The metaphors we've used in the past for the brain have not stood the test of time. I doubt that our current metaphor of the brain as a network of computers doing computations is going to stand for all eternity either.

Note that I do not doubt that there are mechanistic explanations for how we think, and I certainly proceed with my work of trying to build intelligent robots using computation as a primary tool for expressing mechanisms within those robots.

But I have relatively recently come to question computation as the ultimate metaphor to be used in both the understanding of living systems and as the only important design tool for engineering intelligent artifacts.

Some of my colleagues have managed to recast Pluto's orbital behavior as the body itself carrying out computations on forces that apply to it. I think we're perhaps better off using Newtonian mechanics (with a little Einstein thrown in) to understand and predict the orbits of planets and other bodies. It's so much simpler.

Likewise, we can think about spike trains as codes and worry about neural coding. We can think about human memory as data storage and retrieval. And we can think about walking over rough terrain as computing the optimal place to put down each of our feet. But I suspect that somewhere down the line we are going to come up with better, less computational metaphors. The entities we use for metaphors may be more complex, but the useful ones will lead to simpler explanations.

Just as the notion of computation is only a short step beyond discrete mathematics but opens up vast new territories of questions and technologies, these new metaphors might well be just a few steps beyond where we are now in understanding organizational dynamics but may have rich and far-reaching implications in our abilities to understand the natural world and to engineer new creations.

WILLIAM H. CALVIN

Affiliate professor of psychiatry and behavioral sciences emeritus, University of Washington School of Medicine; author of Global Fever: How to Treat Climate Change

Greenland Changed My Mind

Fifty years ago, when I first heard about global warming, almost everyone thought that serious climate problems were several centuries in the future. That's because no one realized how ravenous the world's appetite for fossil fuels would become. The world population is up threefold since then, but fossil fuel consumption is up fourfold, causing problems to arrive on the fast track.

We also thought that problems would develop gradually, allowing for gradual solutions. Wrong again. I tuned in to abrupt climate change about 1984, when the Greenland ice core records of ancient climate showed big jumps in temperature and snowfall, stepping up and down in a mere decade but lasting centuries. I worried about global warming setting off another flip when some tipping point was reached, but I still didn't revise my notions about a slow timescale for the present greenhouse warming.

Modern Greenland changed my mind. About 2004, the speedup of the Greenland glaciers made a lot of climate scientists revise their notions about how fast things were changing. When the summer earthquakes associated with glacial movement doubled and then redoubled in a mere ten years, I felt I was standing on shaky ground—that bigger things could happen at any time.

Then I saw the historical data on major floods and fires— steep increases every decade since 1950 and on all continents. That's not trouble moving around; it is called global climate change.

For drought, which had been averaging about 11 percent of the world's land surface at any one time, there was a step up to a new baseline of 22 percent which occurred with the 1982 El Niño. That's not gradual change but an abrupt shift to a new global climate.

You don't need a fever chart or a climate model to know that something big has been happening since 1950. That's just fact, not a matter for opinion. Climate skeptics, while entitled to their opinions about how good the models and Intergovernmental Panel on Climate Change reports are, are not entitled to ignore a dozen other things pointing in the same direction.

But the most sobering realization came when I was going through the Amazon drought data on the big El Niños in 1972, 1982, and 1997. In the last one, we nearly lost two of the world's three major tropical rain forests to fires. If that mega Niño had lasted two years instead of one, we could have seen the atmosphere's excess CO_2 rise 40 percent over a few years, likely producing an even bigger increase in our climate troubles. Furthermore, missing all of those green leaves to remove CO_2 from the air, the annual bump-up of CO_2 concentration would become half again as big. That's like the disaster movie suddenly speeding up.

And we're not even back-paddling as fast as we can; we're just drifting toward the falls, partying on. If I were a student or young parent seeing my future being trashed, I'd be mad as hell. And hell is a pretty good metaphor for where we are heading if we don't get our act together. Quickly.

Human genome decoder; president and chairman of the J. Craig Venter Institute; author of A Life Decoded; *CEO of Synthetic Genomics*

In Denial

Like many, or perhaps most, I wanted to believe that our oceans and atmosphere were basically unlimited sinks with an endless capacity to absorb the waste products of human existence. I wanted to believe that solving the carbon-fuel problem was for future generations and that the big concern was the limited supply of oil, not the rate of adding carbon to the atmosphere. The data, however, is irrefutable: Carbon dioxide concentrations have been steadily increasing in our atmosphere as a result of human activity since the earliest measurements began. We know that on the order of 4.1 billion tons of carbon are being added to and are staying in our atmosphere each year. We know that burning fossil fuels and deforestation are the principal contributors to the increasing carbon dioxide concentrations in our atmosphere. Eleven of the past twelve years rank among the warmest years since 1850. While no one knows for certain the consequences of this continuing unchecked warming, some have argued that it could result in catastrophic changes, such as the disruption of the Gulf Stream, which keeps the U.K. out of the Ice Age, or even the possibility of the Greenland ice sheet sliding into the Atlantic Ocean. Whether or not these devastating changes occur, we are conducting a dangerous experiment with our planet. One we need to stop.

The developed world, including the United States, England, and Europe, contribute disproportionately to the environmental carbon, but the developing world is rapidly catching up. As the world population increases from 6.5 billion people to 9 billion over the next forty-five years and countries such as India and China continue to industrialize, some estimates indicate that we will be adding more than twenty billion tons of carbon a year to the atmosphere. Continued greenhouse gas emissions at or above current rates would cause further warming and induce many changes to the global climate that could be more extreme than those observed to date. This means we can expect more climate change, more melting ice caps, rising sea levels, warmer oceans and therefore greater storms, and more droughts and floods, all of which compromise food and freshwater production.

It required close to a hundred thousand years for the human population to reach a billion people in 1804. I was born in 1946, when there were only about 2.4 billion of us on the planet. Today there are almost three people for each one of us in 1946 and there will soon be four.

Our planet is in crisis, and we need to mobilize all our intellectual forces to save it. One solution could lie in building a scientifically literate society in order to survive. There are those who like to believe that the future of life on Earth will continue as it has in the past. Unfortunately for humanity, the natural world around us does not care what we believe. But believing we can do something to change our situation, using our knowledge, can very much affect the environment in which we live.

LAURENCE C. SMITH

Professor of geography, University of California, Los Angeles

Faster Than We Thought

The year 2007 marked three memorable events in climate science: release of the Fourth Assessment Report of the Intergovernmental Panel on Climate Change (IPCC AR4); a decade of drought in the American West and the arrival of severe drought in the American Southeast; and the disappearance of nearly half the polar sea ice floating over the Arctic Ocean. The IPCC report (a three-volume, three-thousand-page synthesis of current scientific knowledge written for policymakers) and the American droughts merely hardened my conviction that anthropogenic climate warming is real and just getting going—a view shared (regarding the IPCC report) by the Nobel Foundation. The sea-ice collapse, however, changed my mind that it will be decades before we see the real effects of the warming. I now believe they will happen much sooner.

Let's put the 2007 sea-ice year into context. In the 1970s, when NASA first began mapping sea ice from microwave satellites, its annual minimum extent in September, at summer's end, hovered close to 8 million square kilometers—about the area of the conterminous United States minus Ohio. In September 2007, it dropped abruptly to 4.3 million square kilometers, the area of the conterminous United State minus Ohio and all the other twenty-four states east of the Mississippi, as well as North Dakota, Minnesota, Missouri, Arkansas, Louisiana, and Iowa. Canada's Northwest Passage was freed of ice for the first time in human memory. From Bering Strait, where the United States and Russia brush lips, open blue water stretched almost to the North Pole.

What makes the 2007 sea-ice collapse so unnerving is that it happened too soon. The ensemble averages of our most sophisticated climate-model predictions, put forth in the IPCC AR4 report and various other model intercomparison studies, don't predict a downward lurch of that magnitude for another fifty years. Even the aggressive models—the National Center for Atmospheric Research (NCAR) CCSM3 and the Centre National de Recherches Météorologiques (CNRM) CM3 simulations, for example—must whittle ice until 2035 or later before the 2007 conditions can be replicated. Put simply, the models are too slow to match reality. Geophysicists, accustomed to nonlinearities and hard to impress after a decade of "unprecedented" events, are stunned by the totter: Apparently, the climate system can move even faster than we thought. This has decidedly recalibrated scientists' attitudes, including my own, to the possibility that even the direst IPCC scenario predictions for the end of this century—ten- to twenty-four-inch-higher global sea levels, for example—may be prudish.

What does all this tell us about the future? First, that rapid climate change—a nonlinearity that occurs when a climate forcing reaches a threshold beyond which little additional forcing is needed to trigger a large effect—is a distinct threat not well captured in our current generation of computer models. Models will doubtless improve as the underlying physics of the 2007 ice event and others such as the American Southeast drought are dissected, understood, and codified, but in the meantime policymakers must work from the IPCC blueprint, which seems almost staid after the events of this summer and fall. Second, it now seems probable that the Northern Hemisphere will lose its ice lid far sooner than we thought possible. Over the past three years, experts have shifted from 2050 to 2035 to 2013 as plausible dates for an ice-free Arctic Ocean—estimates at first guided by models, then revised by reality.

The broader significance of vanishing sea ice extends far beyond suffering polar bears, new shipping routes, or even development of vast Arctic energy reserves. It is absolutely unequivocal that the disappearance of summer sea ice—regardless of exactly which year it arrives—will profoundly alter the Northern Hemisphere climate, particularly through amplified winter warming of at least twice the global average rate. Its further influence on the world's precipitation and pressure systems are under study but are likely significant. Effects both positive and negative— from reduced heating-oil consumption to outbreaks of fire and disease—will propagate far southward, into the United States, Canada, Russia, and Scandinavia. Scientists have expected such things to happen eventually, but in 2007 we learned they may already be upon us.

LEE M. SILVER

Professor, Department of Molecular Biology and the Woodrow Wilson School of Public and International Affairs, Princeton University; author of Challenging Nature: The Clash of Science and Spirituality at the New Frontiers of Life

Irrationality and Human Nature

In an interview with the *New York Times* shortly before he died, Francis Crick told a reporter, "The view of ourselves as [ensouled] 'persons' is just as erroneous as the view that the Sun goes around the Earth. This sort of language will disappear in a few hundred years. In the fullness of time, educated people will believe there is no soul independent of the body, and hence no life after death."

Like the vast majority of academic scientists and philosophers alive today, I accept Crick's philosophical assertion—that when your body dies, you cease to exist—without any reservations. I also used to agree with Crick's psychosocial prognosis—that modern education would inevitably give rise to a populace that rejected the idea of a supernatural soul. But on this point, I have changed my mind.

Underlying Crick's psychosocial claim is a common assumption: The minds of all intelligent people must operate according

144

to the same universal principles of human nature. Anyone who makes this assumption will naturally believe that his or her own mind-type is the universal one; in the case of Crick and most other molecular biologists, the assumed universal mind-type is highly receptive to the persuasive power of pure logic and rational analysis.

Once upon a time, my own worldview was similarly informed. I was convinced that scientific facts and rational argument alone could win the day with people who were sufficiently intelligent and educated. To my mind, the rejection of rational thought by such people was a sign of disingenuousness, in service of political or ideological goals.

My mind began to change one evening in November 2003. I had given a lecture at a small liberal arts college, along with a member of the President's Council on Bioethics, whose views on human-embryo research were diametrically opposed to my own. Surrounded by students at the wine-and-cheese reception following our lectures, the two of us began an informal debate about the true meaning and significance of changes in gene expression and DNA methylation during embryonic development. Six hours later, long after the last student had crept off to sleep, it was 4:00 a.m. and we were both still convinced that with just one more round of debate, we'd get the other to capitulate. It didn't happen.

Since this experience, I have purposely engaged other well-educated defenders of the irrational, as well as numerous students at my university, in spontaneous one-on-one debates about a host of contentious biological subjects, including evolution, organic farming, homeopathy, cloned animals, "chemicals" in our food, and genetic engineering. Much to my chagrin, even after politics, ideology, economics, and other cultural issues have been put aside, there is often a refusal to accept scientific implications of rational argumentation.

While its mode of expression may change over cultures and time, irrationality and mysticism seem to be an integral part of normal human nature, even among highly educated people. No matter what scientific and technological advances are made in the future, I now doubt that supernatural beliefs will ever be eradicated from the human species.

LEE SMOLIN

Physicist, Perimeter Institute; author of The Trouble with Physics

About Time

Although I have changed my mind about several ideas and theories, my longest struggle has been with the concept of time.

The most obvious and universal aspect about reality, as we experience it, is that it is structured as a succession of moments, each of which comes into being, supplanting what was just present and is now past. But, as soon as we describe nature in terms of mathematical equations, the present moment and the flow of time seem to disappear, and time becomes just a number, a reading on an instrument, like any other.

Consequently, many philosophers and physicists argue that time is an illusion, that reality consists of the whole four-dimensional history of the universe, as represented in Einstein's theory of general relativity. Some, like Julian Barbour, go further and argue that when quantum theory is unified with gravity, time disappears completely. The world is just a vast collection of moments represented by the "wave-function of the universe." Time not being real, it is just an "emergent quantity" that is helpful in organizing our observations of the universe when it is big and complex.

Other physicists argue that aspects of time are real, such as the relationships of causality that record which events were the necessary causes of others. Roger Penrose, Rafael Sorkin, and Fotini Markopoulou have proposed models of quantum space-time in which everything real reduces to these relationships of causality.

In my own thinking, I first embraced the view that quantum reality is timeless. In our work on loop quantum gravity, we were able to take this idea more seriously than people before us could, because we could construct and study exact wave-functions of the universe. Carlo Rovelli, Bianca Dittrich, and others worked out in detail how time would "emerge" from the study of the question of what quantities of the theory are observable.

But somehow the more this view was worked out in detail, the less I was convinced. This was partly due to technical challenges in realizing the emergence of time and partly because some naive part of me could never understand conceptually how the basic experience of the passage of time could emerge from a world without time.

So in the late '90s I embraced the view that time, as causality, is real. This fit best the next stage of development of loop quantum gravity, which was based on quantum spacetime histories. However, even as we continued to make progress on the technical side of these studies, I found myself worrying that the present moment and the flow of time were still nowhere represented. And I had another motivation, which was to make sense of the idea that laws of nature could evolve in time.

Back in the early 1990s I had formulated a view of laws evolving on a landscape of theories along with the universe they govern. This had been initially ignored, but in the last few years there has been much study of dynamics on landscapes of theories. Most of these are framed in the timeless language of the "wave-function of the universe," in contrast to my original presentation, in which theories evolved in real time. As these studies progressed, it became clear that only those in which time played a role could generate testable predictions—and this made me want to think more deeply about time.

It is becoming clear to me that the mystery of the nature of time is connected with other fundamental questions, such as the nature of truth in mathematics and whether there must be timeless laws of nature.* Rather than being an illusion, time may be the only aspect of our present understanding of nature that is not temporary and emergent.

---

* Recently I have been thinking through this issue again, stimulated by a collaboration with Roberto Mangabeira Unger.

STEPHON ALEXANDER

Assistant professor of physics, Pennsylvania State University

Local or Nonlocal?

Before I entered the intellectual funnel of graduate school, I used to cook up thought experiments to explain coincidences, such as running into someone right after experiencing a random thought about them. This secretive thinking was good mental entertainment, but the demands of forging a serious career in physics research forced me to make peace with such wild speculations. In my theory of coincidences, nonlocal interactions and a dark form of energy were necessary—absolute science fiction! Fifteen years later, we now have overwhelming evidence of a "fifth force," mediating an invisible substance that the physics community has dubbed "dark energy." In hindsight, it is no coincidence that I have changed my mind and now believe that nature is nonlocal.

Nonlocal correlations are not common experiences to us, thus they are difficult both to imagine and accept. Often, research in theoretical physics encourages me to keep an open mind, to not get too attached to ideas that I'm deluded into thinking should be correct. While it's been a constant struggle for me in my scientific career thus far, I have experienced the value of this theoretical ideology-weaning process. After years of wrestling with some of the outstanding problems in the fields of elementary particle physics and cosmology, I have been forced to change my mind on this predisposition silently passed on to me by my physics predecessors: that the laws of physics are for the most part local.

During my first year in graduate school, I came across the famous Einstein-Podolsky-Rosen (EPR) thought experiment, which succinctly argues for "spooky action at a distance" in quantum mechanics. Then came Alain Aspect's experiment that measured the nonlocal entanglement of photon polarization, confirming that there exist nonlocal correlations in nature enabled by quantum mechanics (with a caveat, of course).

This piece of knowledge had a short life in my education and research career. Nonlocality exited the door of my brain after I approached one of my professors, an accomplished quantum field theorist. He convinced me that nonlocality goes away once quantum mechanics properly incorporates causality through a unification with special relativity—a theory known as quantum field theory. With the promise of a sounder career path, I welcomed these then-comforting words and attempted to master quantum field theory. Moreover, even if nonlocality happened, these processes would be exceptional events created under special conditions, whereas most physics was completely local. Quantum field theory works, and it became my new religion.

Now that I specialize in the physics of the early universe, I have witnessed firsthand the great predictive and precise explanatory powers of Einstein's general relativity married with quantum field theory to explain the complete history and physical mechanism for the origin of structures in the universe, all in a seemingly local and causative fashion. We call this paradigm cosmic inflation, and it is deceptively simple. The universe started out immediately after the Big Bang from a microscopically tiny piece of space, then inflated faster than the speed of light. Inflation is able to explain our entire complexity of observed universe with the economy of a few equations that involve general relativity and quantum field theory.

Despite its great success, inflation has been plagued with conceptual and technical problems. These problems created thesis projects and, inevitably, jobs for a number of young theorists like myself. Time after time, publication after publication, like a rat on his wheel, we are running out of steam as the problems of inflation reappear in one form or another. I have now convinced myself that the problems associated with inflation won't go away unless we somehow include nonlocality.

Ironically, inflation is ignited by the same form of dark energy we see permeating the fabric of the cosmos today, except that it was in much greater abundance fourteen billion years ago. Where did most of the dark energy go, after inflation ended? Why is some of it still around? Is this omnipresent dark energy the culprit behind nonlocal activity in physical processes? I don't know exactly how nonlocality in cosmology will play itself out, but by its very nature the physics underlying it will affect "local" processes.

I still haven't changed my mind about coincidences, though.

A. GARRETT LISI

Independent theoretical physicist

We Are Inefficient Inference Engines

As a scientist, I am motivated to build an objective model of reality. Since we always have incomplete information, it is eminently rational to construct a Bayesian network of likelihoods—assigning a probability for each possibility, supported by a chain of priors. When new facts arise, or if new conditional relationships are discovered, these probabilities are adjusted accordingly; our minds should change. When judgment or action is required, it is based on knowledge of these probabilities. This method of logical inference and prediction is the sine qua non of rational thought, and the method all scientists aspire to employ. However, the ambivalence associated with an even probability distribution makes it difficult for an ideal scientist to decide where to go for dinner.

Even though I strive to achieve an impartial assessment of probabilities for the purpose of making predictions, I cannot consider my assessments to be unbiased. In fact, I no longer think humans are naturally inclined to work this way. When I casually consider the beliefs I hold, I cannot readily assign them numerical probabilities. If pressed, I can manufacture these numbers, but this seems more akin to rationalization than rational thought. Also, when I learn something new, I do not immediately erase the information I knew before, even if it is contradictory. Instead, the new model of reality is stacked atop the old. And it is in this sense that a mind doesn't change; vestigial knowledge may fade over a long period of time, but it isn't simply replaced. This model of learning matches a parable from Douglas Adams, relayed by Richard Dawkins:

> A man didn't understand how televisions work and was convinced that there must be lots of little men inside the box, manipulating images at high speed. An engineer explained to him about high-frequency modulations of the electromagnetic spectrum, about transmitters and receivers, about amplifiers and cathode ray tubes, about scan lines moving across and down a phosphorescent screen. The man listened to the engineer with careful attention, nodding his head at every step of the argument. At the end he pronounced himself satisfied. He really did now understand how televisions work. "But I expect there are just a few little men in there, aren't there?"

As humans, we are inefficient inference engines—we are attached to our "little men," some dormant and some active. To a degree, these imperfect probability assessments and pet beliefs provide scientists with the emotional conviction necessary to motivate the hard work of science. Without the hope that an improbable line of research may succeed where others have failed, difficult challenges would go unmet. People should be encouraged to take long shots in science, since, with so many possibilities, the probability of something improbable happening is very high. At the same time, this emotional optimism must be tempered by a rational estimation of the chance of success—

we must not be so optimistic as to delude ourselves. In science, we must test every step, trying to prove our ideas wrong, because nature is merciless. To have a chance of understanding nature, we must challenge our predispositions. And even if we can't fundamentally change our minds, we can acknowledge that others working in science may make progress along their own lines of research. By accommodating a diverse variety of approaches to any existing problem, the scientific community will progress expeditiously in unlocking nature's secrets.

JOHN BAEZ

Mathematical physicist, University of California, Riverside

The String-Loop War

One of the big problems in physics, perhaps the biggest, is figuring out how our two current best theories fit together. On the one hand, we have the Standard Model, which tries to explain all the forces except gravity and takes quantum mechanics into account. On the other hand, we have general relativity, which tries to explain gravity and does *not* take quantum mechanics into account. Both theories seem to be more or less on the right track, but until we somehow fit them together, or completely discard one or both, our picture of the world will be deeply schizophrenic.

It seems plausible that as a step in the right direction we should figure out a theory of gravity that takes quantum mechanics into account but reduces to general relativity when we ignore quantum effects (which should be small in many situations). This is what people mean by "quantum gravity"—the quest for such a theory.

The most popular approach to quantum gravity is string theory. Despite decades of hard work by many very smart people, it's far from clear that this theory is successful. It has made no predictions that have been confirmed by experiment; in fact, it has made few predictions that we have any hope of testing anytime soon. Finding certain sorts of particles at the big new particle accelerator at CERN would count as partial confirmation, but string theory says very little about the details of what we should expect. And thanks to the vast "landscape" of string-theory models that researchers are uncovering, it keeps getting harder to squeeze specific predictions out of this theory.

When I was a postdoc back in the 1980s, I decided I wanted to work on quantum gravity. The appeal of this big puzzle seemed irresistible. String theory was very popular back then, but I was skeptical of it. I became excited when I learned of an alternative approach pioneered by Abhay Ashtekar, Carlo Rovelli, and Lee Smolin called loop quantum gravity. Loop quantum gravity was less ambitious than string theory. Instead of a Theory of Everything, it sought only to be a Theory of Something—namely, a theory of quantum gravity.

So I jumped aboard this train, and for about a decade I was very happy with the progress we were making. A beautiful picture emerged, in which spacetime resembles a random "foam" at very short distance scales, following the laws of quantum mechanics.

We can write down lots of theories of this general sort; however, we have never yet found one for which we can show that general relativity emerges as a good approximation at large distance scales—the quantum soapsuds approximating a smooth surface when viewed from afar, as it were.

I helped my colleagues Dan Christensen and Greg Egan do a lot of computer simulations to study this problem. Most of our results went completely against what everyone had expected. But worse, the more work we did, the more I realized I didn't know what questions we should be asking. It's hard to know what to compute to check that a quantum foam is doing its best to mimic general relativity.

Around this time, string theorists took note of loop quantum gravity people and other critics—in part, thanks to Peter Woit's blog, his book *Not Even Wrong*, and Lee Smolin's *The Trouble with Physics*. String theorists weren't used to criticism like this. A kind of "string-loop war" began. There was a lot of pressure for physicists to take sides for one theory or the other. Tempers ran high. Jaron Lanier put it this way: "One gets the impression that some physicists have gone for so long without any experimental data that might resolve the quantum-gravity debates that they are going a little crazy."

But what was even more depressing was that as this debate raged on, cosmologists were making wonderful discoveries left and right, getting precise data about dark energy, dark matter, and inflation. None of this data could resolve the string-loop war. Why? Because neither of the contending theories could make predictions about the numbers the cosmologists were measuring. Both theories were too flexible.

I realized I didn't have enough confidence in either theory to engage in these heated debates. I also realized there were other questions to work on—questions on which I could actually tell when I was on the right track, questions on which researchers cooperate more and fight less. So I eventually decided to quit working on quantum gravity.

It was very painful to do this, since quantum gravity had been my holy grail for decades. After you've convinced yourself that some problem is the one you want to spend your life working on, it's hard to change your mind. But when I finally did, it was tremendously liberating.

I wouldn't urge anyone else to quit working on quantum gravity. Someday, someone is going to make real progress. When this happens, I may even rejoin the subject. But for now I'm thinking about other things, and I'm making more real progress understanding the universe than I ever did before.

LAWRENCE KRAUSS

Physicist, Arizona State University; author of Hiding in the Mirror: The Quest for Alternate Realities

A Dark Future

Like 99 percent of particle physicists and 60 percent of cosmologists (perhaps 98 percent of the theorists and 90 percent of the observers, to be specific), I was relatively certain there was precisely enough matter in the universe to make it geometrically flat. What does geometrically flat mean? Well, according to general relativity it means there is an exact balance between the positive kinetic energy associated with the expansion of space and the negative potential energy associated with the gravitational attraction of matter in the universe, so that the total energy is zero. This is not only mathematically attractive but in fact is the only theory we have that explains why the universe looks the way it does today.

Now, the only problem with this prediction is that visible matter in the universe accounts for only a few percent of the total amount of matter required to make the universe flat. Happily, however, during the period from 1970 or so to the early 1990s, it became abundantly clear that the galaxies are dominated by dark matter—material that does not shine or, as far as we can tell, interact electromagnetically. This material, which we think is made up of a new type of elementary particle, accounts for at least ten times as much matter as can be accounted for in stars, hot gas, and so on. With the inference that dark matter existed in such profusion, it was natural to suspect that there was enough of it to account for a flat universe.

The only problem was that the more our observations of the universe improved, the less evidence there appeared to be that there was enough dark matter to result in a flat universe. Moreover, all other cosmological indicators, from the age of the universe to the data on large-scale structure, began to suggest that a flat universe dominated by dark matter was inconsistent with observation. In 1995, this led my colleague Mike Turner and I to suggest that the only way a flat universe could be consistent with observation was if most of the energy—indeed almost 75 percent of the total energy—was contributed not by matter but by empty space!

To be fair, I think we were being more provocative than anything else, because the one thing everyone knew was that the energy of empty space had to be precisely zero. The alternative, which would have resulted in something very much like the cosmological constant first proposed by Einstein (when he incorrectly thought the universe was static and needed some adjustment to his equations of general relativity so that the attractive force of gravity was balanced by a repulsive force associated with empty space), was just too ugly to imagine.

Then in 1998, two teams measuring the recession velocity of distant galaxies, using observations of exploding stars within them to probe their distance from us, discovered something amazing. The expansion of the universe seemed to be speeding up with time, not slowing down, as any sensible universe should be doing! Moreover, if one assumed this acceleration was caused by a new repulsive force throughout empty space that would be caused if the energy of empty space was not precisely zero, then the amount of extra energy needed to produce the observed acceleration was precisely the amount needed to account for a flat universe!

Now, here is the really weird thing. Within a year after the observation of an accelerating universe, even though the data was not yet definitive, I and pretty well everyone else in the community who had previously thought there was enough dark matter to result in a flat universe and that the energy of empty space must be precisely zero had completely changed our minds. The signals were just too overwhelming to continue to hold on to our previous rosy picture, even if the alternative was so crazy that none of our fundamental theories could yet account for it.

So we are now pretty sure that the dominant energy-stuff in our universe isn't normal matter and isn't dark matter but rather is associated with empty space. And what is worse (or better, depending on your viewpoint) is that our whole picture of the possible future of the universe has changed. An accelerating universe will carry away almost everything we now see, so that in the far future our galaxy will exist alone in a dark and seemingly endless void.

And that is what I find so satisfying about science. Not just that I could change my own mind because the evidence forced me to, but that the whole community could throw out a cherished notion—and so quickly! That is what makes science different from religion, and that is what makes it worth continuing to ask questions about the universe . . . because it never fails to surprise us.

Psychologist, Harvard University; author of Wet Mind: The New Cognitive Neuroscience

The Environment Sets Up the Brain

I used to believe that we could understand psychology at different levels of analysis, and events at any one of the levels could be studied independently of events at the other levels. For example, one could study events at the level of the brain (and seek answers in terms of biological mechanisms), the level of the person (and seek answers in terms of the contents of thoughts, beliefs, knowledge, and so forth), or the level of the group (and seek answers in terms of social interactions). This approach seemed reasonable; the strategy of "divide and conquer" is a cornerstone in all of science, isn't it? Virtually all introductory psychology textbooks are written as if events at the different levels are largely independent, with separate chapters, only rarely including cross-references to one another, on the brain, perception, memory, personality, social psychology, and so on.

I've changed my mind. I don't think it's possible to understand events at any one level of analysis without taking into account what occurs at other levels. In particular, I'm now convinced that at least some aspects of the structure and function of the brain can be understood only by situating the brain in a specific cultural context. I'm not simply saying that the brain has evolved to function in a specific type of environment, an idea that forms a mainstay of evolutionary psychology and some areas of computer vision, where statistics of the natural environment are used to guide processing. Rather, I'm saying that to understand how any specific brain functions, we need to understand how that person was raised, and currently functions, in the surrounding culture.

Here's my line of reasoning. Let's begin with a fundamental fact: The genes, of which we have perhaps only some thirty thousand, cannot program us to function equally effectively in every possible environment. Hence evolution has licensed the environment to set up and configure each individual's brain so that it can work well in that context. Consider stereovision. We all know about stereo in audition; the sound from each of two loudspeakers has slightly different phases, so the listener's brain glues them together to provide the sense of an auditory panorama. Something similar is at work in vision. In stereovision, the slight disparity in the images that reach the two eyes are a cue for how far away objects are. If you're focused on an object directly in front of you, your eyes will converge slightly. Aside from the exact point of focus, the rest of the image will strike slightly different places on the two retinas (which are at the back of the eye and convert light into neural impulses), and the brain uses the slight disparities to figure out how far away something is.

There are two important points here. First, this stereo process of computing depth on the basis of the disparities where images strike the two retinas depends on the distance between the eyes. And second—and this is absolutely critical—there's no way to know at the moment of conception how far apart a person's eyes are going to be, because that depends on bone growth, and bone growth depends partly on the mother's diet and partly on the infant's diet.

So, given that bone growth depends partly on the environment, how could the genes set up stereovision circuits in the brain? What the genes did is really clever: Young children (peaking at about eighteen months of age) have more connections among neurons than do adults; in fact, until they are about eight years old, children have about twice as many neural connections as they will as adults. But only some of these connections provide useful information. For example, when the infant reaches out, only the connections from some neurons will correctly guide reaching. The brain uses a process called pruning to get rid of the useless connections. The connections that turn out to work, with the distance between the eyes the infant happens to have, would not be the ones that would work if the mother did not have enough calcium or the infant hadn't had enough of various dietary supplements.

This is a really elegant solution to the problem that the genes can't know in advance how far apart the eyes will be. To cope with this problem, the genes overpopulate the brain, giving us options for different environments (where the distance between eyes and length of the arms are part of the brain's "environment," in this sense), and then the environment selects which connections are appropriate. In other words, the genes take advantage of the environment to configure the brain.

This overpopulate-and-select mechanism is not limited to stereovision. In general, the environment sets up the brain (above and beyond any role it may have had in the evolution of the species), configuring it to work well in the world a person inhabits. By "environment," I mean everything outside the brain, including the social environment. For example, it's well known that children can learn multiple languages, without an accent and with good grammar, if they are exposed to the language before puberty. But after puberty, it's very difficult to learn a second language so well. Similarly, when I first went to Japan,

I was told not even to bother trying to bow, that there were something like a dozen different bows and I was always going to "bow with an accent"—and in my case the accent was so thick it was impenetrable.

The notion is that a variety of factors in our environment, including our social environment, configure our brains. It's true for language and I bet it's true for politeness and a raft of other kinds of phenomena. The genes result in a profusion of connections among neurons, which provide a playing field for the world to select and configure so that we fit the environment in which we inhabit. The world comes into our head, configuring us. The brain and its environment are not as separate as they might appear.

This perspective leads me to wonder whether we can assume that the brains of people living in different cultures process information in precisely the same ways. Yes, people the world over have much in common—we are members of the same species, after all—but even small changes in wiring may lead us to use the common machinery in different ways. If so, then people from different cultures may have unique perspectives on common problems and be poised to make unique contributions toward solving such problems.

Changing my mind about the relationship between events at different levels of analysis has led me to change fundamental beliefs. In particular, I now believe that understanding how the surrounding culture affects the brain may be of more than merely academic interest.

ERNST PÖPPEL

Neuroscientist, chairman of the board of the Center for Human Sciences and director of the Institute for Medical Psychology, University of Munich; author of Mindworks: Time and Conscious Experience

The Wittgenstein Straitjacket

When I look at something, when I talk to somebody, when I write a few sentences about "What I have changed my mind about and why," the neuronal network in my brain changes all the time and there are even structural changes in the brain. Why is it that these changes don't usually come to mind but remain subthreshold? Certainly, if everything came to mind that goes on in the brain, and if there were not an efficient mechanism of informational garbage disposal, we would end up in mental chaos (which sometimes happens in unfortunate cases of neuronal dysfunctioning). It is only sometimes that certain events produce so much neuronal energy and attract so much attention that a conscious representation is made possible.

As most neuronal information-processing remains in mental darkness, it is in my view impossible to make a clear statement as to why somebody changed his or her mind about something. If people give an explicit reason for having changed their mind about something, I am suspicious. As these processes are beyond voluntary control, I am much less transparent to myself than I might want, and this is true for everybody. Thus I cannot give a good reason why I changed my mind about a strong hypothesis, or even a belief, or perhaps a prejudice in my scientific work I had until several years ago.

A sentence of Ludwig Wittgenstein's from his *Tractatus Logico-Philosophicus* (5.6) was like a dogma for me: *"Die Grenzen meiner Sprache bedeuten die Grenzen meiner Welt."* ("The limits of my language signify the limits of my world"—my translation.) Now I react to this sentence with an emphatic "No!"

As a neuroscientist, I have to stay away from the language trap. In our research, we are easily misguided by words. Without too much thinking, we refer to "consciousness," to "free will," to "thoughts," to "attention," to "the self," and so on; and we give an ontological status to these terms. Some people even start to look at the potential site of consciousness or of free will in the brain; some people ask the "what is . . ." question that never can find an answer. The prototypical "what is . . ." question was formulated 1,600 years ago by St. Augustine, who said in the eleventh book of his *Confessions*, *"Quid est ergo tempus? Si nemo ex me quaerat scio, si quaerenti explicare velim nescio."* ("What is time? If nobody asks me, I know it, but if I have to explain it to somebody, I don't know it"—my translation.)

Interestingly, Augustine made a nice categorical mistake by referring to "knowing" on an implicit and then on an explicit level. This categorical mistake is still with us when we ask questions like "What is consciousness?" or "What is free will?": One knows, but one does not. As neuroscientists, we have to focus on processes in the brain that rarely, or perhaps never, map directly onto such terms as we use them. Complexity reduction in brains is necessary, and it happens all the time, but the goal of this reductive process is not such terms, which might be useful for communication, but efficient action. This is what I think today, but why I came to this conclusion I don't know. There were probably several reasons that finally resulted in a shift of mind—that is, in overcoming Wittgenstein's straitjacket.

SCOTT D. SAMPSON

Research curator, Utah Museum of Natural History; research associate professor of geology and geophysics, University of Utah; host of Dinosaur Planet

The Coup de Grâce from Space

What killed off the dinosaurs? An asteroid did it. . . .

OK, so this may not seem like news to you. The father-son team of Luis and Walter Alvarez first put forth the asteroid hypothesis in 1980 to account for the extinction of dinosaurs and many other life-forms at the end of the Mesozoic, about 65.5 million years ago. According to this now familiar scenario, an asteroid some ten kilometers in diameter slammed into the planet at about a hundred thousand kilometers per hour. Upon impact, the bolide disintegrated, vaporizing a chunk of the Earth's crust and propelling a gargantuan cloud of gas and dust high into the atmosphere. This airborne matter circulated around the globe, blocking out the sun and halting photosynthesis for a period of weeks or months. As if turning the lights out weren't bad enough, massive wildfires and copious amounts of acid rain apparently ensued.

Put simply, it was hell on Earth. Species succumbed in great numbers, and food webs collapsed the world over, ultimately wiping out about half of the planet's biodiversity. Key geologic evidence includes remnants of the murder weapon; iridium, an element that occurs in small amounts in the Earth's crust but is abundant in asteroids, was found by the Alvarez team to be anomalously abundant in a thin layer within Cretaceous-Tertiary (K-T) boundary sediments at various sites around the world.

In 1990, announcement came of the discovery of the actual impact crater in the Gulf of Mexico. It seemed as though what was arguably the most enduring mystery in prehistory had finally been solved. Unsurprisingly, this hypothesis was also a media darling, providing a tidy yet attractively violent explanation for one of paleontology's most perplexing problems—with the added bonus of a possible repeat performance, this time with humans on the roster of victims.

To some paleontologists, however, the whole idea seemed a bit too tidy.

Ever since the Alvarezes proposed the asteroid (or "impact winter") hypothesis, many if not most dinosaur paleontologists have argued for an alternative scenario to account for the K-T extinction. I was among the ranks of doubters. It's not that I and my colleagues questioned the occurrence of the asteroid impact; supporting evidence for this catastrophic event has been firmly established for some time. At issue is the duration of the event. Whereas the impact hypothesis invokes a rapid extinction—on the order of weeks to years—others argue for a more gradual extinction, one that spanned from one million to several million years. Evidence cited in support of the latter view includes a drop in global sea levels at the end of the Cretaceous and a multimillion-year bout of volcanism. The debate has currently been reduced to two alternatives: the Alvarez scenario, proposing that the K-T extinction was a sudden event triggered by a single extraterrestrial bullet, versus the gradualist view, which proposes that the asteroid impact was accompanied by two other global-scale perturbations—volcanism and decreasing sea level—and that it was only this combination of factors acting in concert that decimated the biosphere at the end of the Mesozoic.

Paleontologists of the gradualist ilk argue that dinosaurs and certain other groups were already on their way out well before the K-T "big bang" occurred. Unfortunately, the fossil record of dinosaurs is relatively poor for the last stage of the Mesozoic, and only one place on Earth—a small swath of badlands in the western interior of North America—has been investigated in detail. Several authors have argued that the latest Cretaceous Hell Creek fauna (as it's called; best known from eastern Montana) is extremely sparse compared with earlier dinosaur faunas. In particular, comparisons have often been made with the circa seventy-five-million-year-old Late Cretaceous Dinosaur Park Formation of southern Alberta, which has yielded a bewildering array of herbivorous and carnivorous dinosaurs.

For a long time, I regarded myself a card-carrying member of the gradualist camp. However, at least two lines of evidence have persuaded me to change my mind and join the ranks of the sudden-extinction-precipitated-by-an-asteroid group.

First is a growing database indicating that the terminal Cretaceous world was not stressed to the breaking point and awaiting arrival of the coup de grâce from outer space. With regard to dinosaurs in particular, recent work has demonstrated that the Hell Creek fauna is much more diverse than previously realized. Second, new and improved stratigraphic age controls for dinosaurs and other Late Cretaceous vertebrates in the Western Interior of the U.S. indicate that ecosystems like those preserved in the Dinosaur Park Formation were not nearly as diverse as previously supposed. Instead, many dinosaur species appear to have existed for relatively short durations (less than a million years), with some geologic units preserving a succession of relatively short-lived faunas. So even within the well-sampled American West (let alone the rest of the world, for which we currently have little hard data), I see no grounds for arguing that dinosaurs were undergoing a slow, attritional demise. Other groups, like plants, also seem to have been doing fine in the interval leading up to that fateful day 65.5 million years ago. Finally, extraordinary events demand extraordinary explanations, and it does not seem parsimonious to make an argument for a lethal cascade of agents when compelling evidence exists for a single agent capable of doing the job on its own.

So yes, as far as I'm concerned (at least for now), the asteroid did it.

Futurist, business strategist; cofounder of the Global Business Network, a Monitor company; author of The Long Boom

The Greater Risk

In the last few years I have changed my mind about nuclear power. I used to believe that expanding nuclear power was too risky. Now I believe that the risks of climate change are much greater than the risks of nuclear power. We need to move urgently toward a new generation of nuclear reactors.

What led to the change of view? I came to believe that the likelihood of major climate-related catastrophes was increasing rapidly and that they were likely to occur much sooner than the simple linear models of the Intergovernmental Panel on Climate Change indicated. My analysis developed as a result of work we did for the defense and intelligence communities on the national security implications of climate change. Many regions of the Earth are likely to experience an increasing frequency of extreme weather events. These catastrophic events include megastorms, supertornados, torrential rains and floods, extended droughts, and ecosystem disruptions, all added to steadily rising sea levels. It also became clear that human-induced climate change is ever more at the causal center of the story.

Research by such climatologists as William Ruddiman indicate that the climate is sensitive to changes wrought by human societies, ranging from such agricultural practices as forest clearing and irrigated rice growing to major plagues to the use of fossil fuels. Human societies have often gone to war as a result of the ecological exhaustion of their local environments—so it becomes an issue of war and peace. Will Vietnam, for instance, simply roll over and die when the Chinese dam what remains of the trickle of the Mekong as an extended drought develops at its source in the Tibetan highlands?

Even allowing for much greater efficiency and a huge expansion of renewable energy, the real fuel of the future is coal, especially in the United States, China, and India. If all three go ahead with their current plans to build coal-fired electric generating plants, that alone will, over the next two decades, double all the CO_2 that humankind has put into the atmosphere since the Industrial Revolution began, more than two hundred years ago.

The only meaningful alternative to coal is nuclear power. We can hope that our ability to capture the CO_2 from coal burning and sequester it in various ways will grow, but it will take a decade or more before that technology reaches commercial maturity.

At the same time, I also came to believe that the risks of nuclear power are less than we feared. That shift began with a trip to visit the proposed nuclear-waste depository at Yucca Mountain in Nevada. When it becomes clear that the very long-term storage of waste (say, for 10,000 to 250,000 years) is a silly idea, not meaningfully realistic, one begins to question many of the assumptions about the future of nuclear power. The right answer to nuclear waste is temporary storage, for perhaps decades, and then recycling of the fuel, as much of the world already does—not sticking it underground for millennia. We will likely need the fuel we can extract from the waste.

There are emerging technologies for both nuclear power and waste reprocessing that will reduce safety risk, the amount of waste, and most especially the risk of nuclear-weapons proliferation as the new fuel cycle produces no plutonium, the substance of concern. And the economics are increasingly favorable, as the French have demonstrated for decades. The average French citizen produces 70 percent less CO_2 than the average American. We have also learned that the long-term consequences of Chernobyl, the worst nuclear accident in history, were much less than feared.

So the conclusion is that the risks of climate change are far greater than the risks of nuclear power. Furthermore, human skill and knowledge in managing a nuclear system are only likely to grow with time—while the risks of climate change will grow as billions more people get rich and change the face of the planet with their demands for more and more stuff. Nuclear power is the only source of electricity we know of that is likely to enable the next three billion or four billion, who will want what we all have, to get what they want without radically changing the Earth's climate.

KEVIN KELLY

Editor at large of Wired; *author of* New Rules for the New Economy

The Collaborative Community

Much of what I believed about human nature, and the nature of knowledge, has been upended by the Wikipedia. I knew that the human propensity for mischief among the young and bored—of which there were many online—would make an encyclopedia editable by anyone an impossibility. I also knew that even among the responsible contributors, the temptation to exaggerate and misremember what we think we know would be inescapable, adding to the impossibility of a reliable text. I knew from my own twenty-year experience online that you cannot rely on what you read in a random posting, and I believed that an aggregation of random contributions would be a total mess. Even unedited Web pages created by experts failed to impress me, so an entire encyclopedia written by unedited amateurs, not to mention ignoramuses, seemed destined to be junk.

Everything I knew about the structure of information convinced me that knowledge would not spontaneously emerge from data without a lot of energy and intelligence deliberately directed to transforming it. All the attempts at headless collective writing I had been involved with had generated only forgettable trash. Why would anything online be any different?

When the first incarnation of the Wikipedia (then called Nupedia) launched in 2000, I gave it a look and was not surprised that it never took off; there was a laborious process of top-down editing and rewriting that discouraged a would-be random contributor. When the back-office wiki created to facilitate the administration of the Nupedia text became the main event and anyone could edit as well as post an article, I expected even less from the effort, now renamed Wikipedia.

How wrong I was. The success of the Wikipedia keeps surpassing my expectations. Despite the flaws of human nature, it keeps getting better. Both the weaknesses and virtues of individuals are transformed into common wealth, with a minimum of rules and elites. It turns out that with the right tools, it is easier to restore damage text (the revert function on Wikipedia) than to create damage text (vandalism) in the first place, and so the good-enough article prospers and continues. With the right tools, the collaborative community can outpace the same number of ambitious individuals competing.

It has always been clear that collectives amplify power—that is what cities and civilizations are—but what's been the big surprise for me is how little the tools and oversight are needed. The bureaucracy of Wikipedia is relatively so small as to be invisible. It's the embedded code-based governance, versus manager-based governance, that is the real news. Yet the greatest surprise brought by the Wikipedia is that we still don't know how far this power can go. We haven't seen the limits of wiki-ized intelligence. Can it make textbooks, music, and movies? What about law and political governance?

Before we say, "Impossible!" I say, "Let's see." I know all the reasons why law can never be written by know-nothing amateurs. But having already changed my mind once on this, I am slow to jump to conclusions again. The Wikipedia is impossible, but here it is. It is one of those things impossible in theory but possible in practice. Once you confront the fact that it works, you have to shift your expectation of what else impossible in theory might work in practice.

I am not the only one who has had his mind changed about this. The reality of a working Wikipedia has made a type of communitarian socialism not only thinkable but desirable. Along with other tools, such as open-source software (and open-source everything), this communitarian bias runs deep in the online world. In other words, it runs deep in the young next generation. It may take several decades for this shifting world perspective to show its full colors. When you grow up knowing (rather than admitting) that such a thing as the Wikipedia works; when it is obvious to you that open-source software is better; when you are certain that sharing your photos and other data yields more than safeguarding them—then these assumptions will become a platform for a yet more radical embrace of the commonwealth. I hate to say it, but there is a new type of communism or socialism loose in the world—although neither of those outdated and tinged terms can accurately capture what is new about it.

The Wikipedia has changed my mind and led me, a fairly steady individualist, toward this new social sphere. I am now much more interested both in the new power of the collective and the new obligations of individuals toward the collective. In addition to expanding civil rights, I want to expand civil duties. I am convinced that the full impact of the Wikipedia is still subterranean, and that its mind-changing power is working subconsciously on the global millennial generation, providing it with proof of a beneficial hive mind and an appreciation for believing in the impossible.

That's what it's done for me.

ALAN KAY

Computer scientist; president of the Viewpoints Research Institute; adjunct professor of computer science, University of California, Los Angeles

Non-Story Thinking

When I was age ten, in 1950, I first encountered the department-store pneumatic-tube system for moving receipts and money from the store's counters to the cashier's office. I loved this and tried to figure out how it worked. The clerks in the store knew all about it. "Vacuum," they said. "Vacuum sucks the canisters. Just like your mom's vacuum cleaner."

"But how does it work?" I asked.

"Vacuum," they said. "Vacuum does it all." This was what adults called "an explanation."

So I took apart my mom's Hoover to find out how it worked. There was an electric motor in there, which I had expected, but the only other thing in there was a fan. How could a fan produce a vacuum, and how could it suck?

We had a room fan, and I looked at it more closely. I knew that it worked like the propeller of an airplane, but I'd never thought about how those worked. I picked up a board and moved it. It moved air just fine. So the blades of the propeller and the

179

fan were just boards that the motor kept on moving, in order to push air.

But what about the Hoover? I found that a sheet of paper would stick to the back of the fan. But why? I "knew" that air was supposed to be made up of particles too small to be seen. So it was clear why you got a gust of breeze by moving a board: You were knocking little particles one way and not another. But where did the sucking of the paper on the fan and the sucking in the vacuum cleaner come from?

Suddenly it occurred to me that the air particles must already be moving very quickly and bumping into one another. When the board, or fan blades, moved air particles away from the fan there were less near the fan and the already moving particles would have less to bump into and would thus move toward the fan. They didn't know about the fan, but they appeared to.

The "suck" of the vacuum cleaner was not a suck at all. What was happening was that things went into the vacuum cleaner because they were being "blown in" by the air particles' normal movement, which was not being opposed by the usual pressure of air particles inside the fan.

When my physiologist father came home that evening, I exclaimed, "Dad, the air particles must be moving at least a hundred miles an hour!" I told him what I'd found out, and he looked in his physics book. In there was a formula to compute the speed of air molecules at various temperatures. It turned out that at room temperature, ordinary air molecules were moving much faster than I had guessed—more like 1,500 miles an hour. This completely blew my mind.

Then I got worried, because even small things were clearly not moving that fast going into the vacuum cleaner (or inside the pneumatic tubes). By putting my hand out the window of the car, I could feel that the air was probably going into the vacuum cleaner closer to fifty or sixty miles an hour. Another conversation with Dad led to two ideas: (a) the fan was probably not very efficient at moving particles away, and (b) the particles themselves were going in every direction and bumping into one another (this is why it takes a while for perfume from an open bottle to be smelled across a room).

This experience was a big deal for me, because I had thought one way, using a metaphor and a story about "sucking," and then I suddenly thought just the opposite, because of an experiment and non-story thinking. The world was not as it seemed—or as most adults thought and claimed! I never trusted a "story" again.

DIANE F. HALPERN

Professor of psychology, Claremont McKenna College; past president of the American Psychological Association; author of Sex Differences in Cognitive Abilities

Establishing Cognitive Sex Differences

Why are men underrepresented in teaching, child care, and related fields and women underrepresented in engineering, physics, and related fields? I used to know the answer, but that was before I spent several decades reviewing almost everything written about this question. Like most enduring questions, the responses have grown more contentious and even less is settled now that we have mountains of research designed to answer them. At some point, my own answer changed from what I believed to be the simple truth to a convoluted statement complete with qualifiers, hedge terms, and caveats. I guess this shift in my own thinking represents progress, but it doesn't feel or look that way.

I am a feminist, a product of the 1960s, someone who believed that group differences in intelligence or almost any other trait are mostly traceable to the lifetime of experiences that mold us. Of course, I never doubted the basic premises of evo-lution, but the lessons I learned from evolution favor the idea that brain and behavior are adaptable. Hunter-gatherers never solved calculus problems or traveled to the moon, so little in our ancient past explains these modern-day achievements.

There is also the disturbing fact that evolutionary theories can easily explain almost any outcome, so I never found them to be a useful framework for understanding behavior. Even when I knew the simple truth about sex differences in cognitive abili-ties, I never doubted that heritability plays a role in cognitive development—but, like many others, I believed that once the potential to develop an ability exceeded some threshold value, heritability was of little importance. Now I am less sure about any single answer, and nothing is simple anymore.

The literature on sex differences in cognitive abilities is filled with inconsistent findings, contradictory theories, and emotional claims unsupported by the research. Yet despite all the noise in the data, clear and consistent messages can be heard. There are real, and in some cases sizable, sex differences with respect to some cognitive abilities.

Socialization practices are undoubtedly important, but there is also good evidence that biological sex differences play a role in establishing and maintaining cognitive sex differences, a conclusion I wasn't prepared to make when I began review-ing the relevant literature. I could not ignore, or explain away, repeated findings about (small) variations over the menstrual cycle; the effects of exogenously administered sex hormones on cognition; a variety of anomalies that allow us to separate prenatal hormone effects on later development; failed attempts to alter the sex roles of a biological male after an accident had destroyed his penis; differences in preferred modes of thought; international data on the achievement of females and males— to name just a few types of evidence that demand the conclusion that there is some biological basis for sex-typed cognitive development.

My thinking about this controversial topic has changed. I have come to understand that nature needs nurture and the dichotomization of these two influences on development is the wrong way to conceptualize their mutual influences on each other. Our brain structures and functions reflect and direct our life experiences, which create feedback loops that alter the hormones we secrete and how we select environments. Learning is a biological and environmental phenomenon.

What had been a simple truth morphed into a complicated answer for the deceptively simple question of why there are sex differences in cognitive abilities. There is nothing in my new understanding that justifies discrimination or predicts the continuation of the status quo. There is plenty of room for motivation, self-regulation, and persistence to make the question about the underrepresentation of women and men in different academic areas moot in coming years.

Like all complex questions, the question about why men and women achieve in different academic areas depends on a laundry list of influences that do not fall neatly into categories labeled "biology" or "environment." It is time to give up this tired way of thinking about nature and nurture as two independent variables and recognize that they exert influences on each other. No single number can capture the extent to which one type of variable is important, because they do not operate independently. Nature and nurture do not just interact; they fundamentally change each other. The answer I give today is far more complicated than the simple truth I used to believe, but we have no reason to expect that complex phenomena like cognitive development have simple answers.

STEPHEN H. SCHNEIDER

Biologist and climatologist, Stanford University; author of Laboratory Earth

What Responsible Scientists Must Do When the Facts Change

In my public talks on global warming, even these days, I often hear, "I don't believe in global warming," and I then typically get asked why I do, "when all the evidence is not in."

"Global warming is not about beliefs but an accumulation of evidence over decades," I typically retort, "so that we can now say that the vast preponderance of evidence and its consistency with basic climate theory supports global warming as well established—not that all aspects are fully known, which is an impossibility in any complex-systems science."

But it hasn't always been that way—especially for me, at the outset of my career in 1971, when I coauthored a controversial paper calculating that cooling effects from a shroud of atmospheric dust and smoke (aerosols) from human emissions on a global scale appeared to dominate the opposing warming effect of the growing atmospheric concentrations of the greenhouse gas carbon dioxide. Measurements at the time showed that both

warming and cooling emissions were on the rise, so a calculation of the net balance was essential, though controlling the
aerosols made sense with or without climate side effects, since
they posed—and still pose—serious health effects on vulnerable
populations. In fact, for the latter reason, laws to clean up the air
in most rich countries were being negotiated at about that time.

When I traveled the globe in the early 1970s to explain our
calculations, what I slowly learned from those out there making measurements was that two facts had only recently come to
light; together, they made me consider flipping sign from cooling to warming as the most likely climatic-change direction from
humans using the atmosphere as a free sewer to dump some of our
volatile industrial and agricultural wastes. These facts were that
human-injected aerosols, which we assumed were global in scale
in our cooling calculation, were in fact concentrated primarily in
industrial regions and biomass-burning areas of the globe—about
20 percent of the Earth's surface—whereas we already knew that
CO_2 emissions were global in extent and about half of the emitted
CO_2 would last for a century or more in the air.

But there was something even more convincing: CO_2 was not
the only important human-emitted greenhouse gas; there were also
methane, nitrous oxide, and chlorofluorocarbons (many of this last
type now banned, because chlorofluorocarbons also deplete strato-
spheric ozone), and together with CO_2, these made for an enhanced
global set of warming factors. Aerosols, on the other hand, were primarily regional and thus could not overcome the warming effects of
the combined global-scale greenhouse gases.

I am very proud to have published, in the mid-1970s, what
was wrong with my early calculations, well before the so-called
"contrarians"—climate-change deniers still all too prevalent even
today—understood the issues, let alone incorporated the new
facts into updated models to make more credible projections. Of
course, today the dominance of warming over cooling agents is

well established in the climatology community, but our remaining inability to be precise over how much warming the planet will have to deal with is largely still due to uncertainty over the partially counteracting cooling effects of aerosols—enough to offset a significant, even if unknown, amount of warming. So although we are confident of the existence of human-caused warming in the past several decades from greenhouse gases, we are still working hard to pin down much more precisely how much aerosols offset this warming. Facts on that offset still lag behind the critical need to better estimate our effects on climate before they become potentially irreversible.

The sad part of this story is not about science but the misinterpretation of it in the political world. I still have to endure polemical blogs from contrarian columnists and others about how, as one of them put it, "Schneider is an environmentalist for all temperatures"—citing my early calculations. This famous columnist somehow forgot to bring up my later correction of my faulty assumptions or mention that the 1971 calculation was based on not-yet-gathered facts. Simply getting the sign wrong was cited, ipso facto in this blog, as destructive of my current credibility.

Ironically, inside the scientific world this switch of sign of projected effects is viewed as exactly what responsible scientists must do when the facts change. Not only did I change my mind but I published almost immediately what had changed and how that played out over time. Scientists have no crystal ball, but we do have modeling methods that are the closest approximation available. They can't give us truth, but they can tell us the logical consequences of explicit assumptions. Those who update their conclusions explicitly as facts evolve are much more likely to be a credible source than those who stick to old stories for political consistency. Two cheers for the scientific method!

Tech-culture journalist; coeditor of the blog Boing Boing; commentator on NPR; host of Boing Boing tv

Noise on the Blog

I changed my mind about online community this year.

I coedit a blog that attracts a large number of daily visitors, many of whom have something to say back to us about whatever we write or produce in video. When our audience was small, in the early days, interacting was simple: We tacked a little href tag to an open-comments thread at the end of each post: Link, Discuss. No moderation, no complication, come as you are, anonymity's fine. Every once in a while, a thread accumulated more noise than signal, but the balance mostly worked.

But then the audience grew. Fast. And with that grew the number of antisocial actors, "drive-by trolls," people for whom dialogue wasn't the point. It doesn't take many of them to ruin the experience for the much larger numbers of participants who are acting in good faith.

Some of the more grotesque attacks were aimed at me, and the new experience of being on the receiving end of that much personally directed nastiness was upsetting. I dreaded hitting the "publish" button on posts, because I knew what now would follow.

The noise on the blog grew, the interaction ceased to be fun for anyone, and with much regret we removed the comments feature entirely.

I grew to believe that the easier it is to post a drive-by comment—and the easier it is to remain faceless, reputation-less, and real-world-less while doing so—the greater the volume of antisocial behavior that follows. I decided that no online community could remain civil after it grew too large, and I gave up on that aspect of Internet life.

My coeditors and I debated, we brainstormed, we observed other big sites that included some kind of community forum or comments feature. Some relied on voting systems to "score" whether a comment is of value; this felt clinical, cold, like grading what a friend says to you in conversation. Dialogue shouldn't be a beauty contest. Other sites used other automated systems to rank the relevance of a speech thread. None of this felt natural to us or seemed an effective way to prevent the toxic-sludge buildup. So we stalled for years, and our blog remained more monologue than dialogue. That felt unnatural, too.

Finally, this year, we resurrected comments on the blog, with the one thing that did feel natural. Human hands.

We hired a community manager and equipped our comments system with a secret weapon: the "disemvoweller." If someone is misbehaving, she can remove all the vowels from their screed with one click. The dialogue stays but the misanthrope looks ridiculous, and the emotional sting is neutralized.

Now, once again, the balance mostly works. I still believe that there is no fully automated system capable of managing the complexities of online human interaction—no software fix I know of. But I'd underestimated the power of dedicated human attention.

Plucking one early weed from a bed of germinating seeds changes everything. Small actions by focused participants change the tone of the whole. It is possible to maintain big healthy gardens online. The solution isn't cheap, or easy, or hands-free. Few things of value are.

SHERRY TURKLE

Psychologist, Massachusetts Institute of Technology; author of Evocative Objects: Things We Think With

The Robot in the Wings

Throughout my academic career—when I was studying the relationship between psychoanalysis and society and when I moved to the social and psychological studies of technology— I've seen myself as a cultural critic. I don't mention this to stress how lofty a job I put myself in but rather to emphasize that I saw the job as theoretical in its essence. Technologists design things; I was able to offer insights about the nature of people's connections to them, the mix of feelings in the thoughts, how passions mixed with cognition. Trained in psychoanalysis, I didn't see my stance as therapeutic, but it did borrow from the reticence of that discipline. I was not there to meddle, I was there to listen and interpret. Over the past year, I've changed my mind: Our current relationship with technology calls forth a more meddlesome me.

In the past, because I didn't criticize but tried to analyze, some of my colleagues found me complicit with the agenda of technology builders. I didn't like that much, but I understood that this was perhaps the price to pay for maintaining my distance, as Little Red Riding Hood's wolf would say, "the better to hear them with." This year I realized I had changed my stance. In studying reactions to advanced robots—robots that look you in the eye, remember your name, and track your motions—I found more and more people who were considering such robots as friends, confidants, and (as they imagined technical improvements) even as lovers. I became less distanced. I began to think about technological promiscuity. Are we so lonely that we love whatever is put in front of us?

I kept listening for what stood behind the new promiscuity (my habit of listening didn't change), and I began to get evidence of a certain fatigue with the difficulties of dealing with people. A female graduate student came up to me after a lecture and told me she would gladly trade in her boyfriend for a sophisticated humanoid robot as long as the robot could produce what she called "caring behavior." She told me that she needed "the feeling of civility in the house and I don't want to be alone." She said, "If the robot could provide a civil environment, I would be happy to help produce the illusion that there is somebody really with me." What she was looking for, she told me, was a "no-risk relationship" that would stave off loneliness; a responsive robot, even if it was just exhibiting scripted behavior, seemed better to her than a demanding boyfriend. I thought she was joking. She was not.

In a way, I should not have been surprised. For a decade, I had studied the appeal of sociable robots. They push our Darwinian buttons. They are programmed to exhibit the kind of behavior we have come to associate with sentience and empathy, which leads us to think of them as autonomous creatures with intentions and emotions. Once people see robots as creatures, they feel a desire to nurture them. With this feeling comes the fantasy of reciprocation. As you begin to care for these creatures, you want them to care about you.

And yet, in the past, I had found that people approached computational intelligence with a certain "romantic reaction."

Their basic position was that simulated thinking might be thinking but simulated feeling was never feeling and simulated love was never love. Now I was hearing something new. People were more likely to tell me that human beings might be "simulating" their feelings; as one woman put it, "How do I know that my lover is not just simulating everything he says he feels?" Everyone I spoke with was busier than ever with their e-mail and their virtual friendships. Everyone was busier than ever with their social networking and always-on/always-on-you PDAs. Someone once said that loneliness is failed solitude. Could no one stand to be alone anymore, before they turned to a device? Were cyberconnections paving the way to thinking that a robot might be sufficient unto the day? I was left contemplating not the cleverness of engineering but the vulnerabilities of people.

Last spring, I had a public exchange in which a colleague wrote about the "I-Thou" dyad of people and robots, and I could sense Martin Buber spinning in his grave. The "I" was the person in the relationship, but how could the robot be the "Thou"? I once would have approached such an interchange with discipline, interested only in the projection of feeling onto the robot. But I had taken that position when robots seemed only an evocative object for better understanding people's hopes and frustrations. Now people were doing more than fantasizing. There was a new earnestness. They saw the robot in the wings and were excited to welcome it onstage.

In no time at all, it seemed, a book came out called *Love and Sex with Robots* and a reporter from *Scientific American* was interviewing me about the psychology of robot marriage. The conversation was memorable, and I warned my interviewer that I would use it as data. He asked me if my opposition to people marrying robots put me in the same camp as those who oppose gay marriage. I tried to explain that just because I didn't think people could marry machines didn't mean I didn't think mixes of people with people was fair play. He accused me of species chauvinism. Wasn't this the kind of talk that homophobes once used—not considering gays as "real people"? Right there, I changed my mind about my vocation. I changed my mind about where my energies were most needed. I was turning in my card as a cultural critic the way I had always envisaged that identity. Now I was a cultural *critic*. I wasn't neutral, I was very sad.

DANIEL GILBERT

Professor of psychology, Harvard University; author of Stumbling on Happiness

Happy with What You've Got

Six years ago, I changed my mind about the benefit of being able to change my mind.

In 2002, Jane Ebert and I discovered that people are generally happier with decisions when they can't undo them. When subjects in our experiments were able to undo their decisions they tended to consider both the positive and negative features of the decisions they had made, but when they couldn't undo their decisions they tended to concentrate on the good features and ignore the bad. Thus they were more satisfied when they made irrevocable decisions than when they made revocable ones. Ironically, subjects did not realize this would be the case and strongly preferred having the opportunity to change their minds.

Now, up until this point I had always believed that love causes marriage. But these experiments suggested to me that marriage could also cause love. If you take data seriously, you act on it, so when these results came in, I went home and proposed to the woman I was living with. She said yes, and it turned out that the data were right: I love my wife more than I loved my girlfriend.

The willingness to change one's mind is a sign of intelligence, but the freedom to do so comes at a cost.

*Psychologist, Princeton University; recipient of the 2002
Nobel Prize in Economic Sciences*

What Constitutes Life Satisfaction?

The central question for students of well-being is the extent to which people adapt to circumstances. Ten years ago, the generally accepted position was that there is considerable hedonic adaptation to life conditions. The effects of circumstances on life satisfaction appeared surprisingly small: The rich were only slightly more satisfied with their lives than the poor, the married were happier than the unmarried but not by much, and neither age nor moderately poor health diminished life satisfaction. Evidence that people adapt—though not completely—to becoming paraplegic or winning the lottery supported the idea of a "hedonic treadmill": We move, but we remain in place. The famous Easterlin paradox seemed to nail it down: Self-reported life satisfaction has changed very little in prosperous countries over the last fifty years, in spite of large increases in the standard of living.

Hedonic adaptation is a troubling concept, regardless of where you stand on the political spectrum. If you believe that economic growth is the key to increased well-being, the Easterlin paradox is bad news. If you are a compassionate liberal, the finding that the sick and the poor are not very miserable takes wind from your sails. And if you hope to use a measure of well-being to guide social policy, you need an index that will pick up permanent effects of good policies on the happiness of the population.

About ten years ago, I had an idea that seemed to solve these difficulties: Perhaps people's satisfaction with their life is not the right measure of well-being. The idea took shape in discussions with my wife, Anne Treisman, who was (and remains) convinced that people are happier in California (or at least Northern California) than in most other places. The evidence showed that Californians are not particularly satisfied with their lives, but Anne was unimpressed. She argued that Californians are accustomed to a pleasant life and come to expect more pleasure than the unfortunate residents of other states. Because they have a high standard for what life should be, Californians are not more satisfied than others, although they are actually happier. This idea included a treadmill, but it was not hedonic, it was an aspiration treadmill: Happy people have high aspirations.

The aspiration treadmill offered an appealing solution to the puzzles of adaptation. It suggested that measures of life satisfaction underestimate the benefits of life circumstances such as income, marital status, or living in California. The hope was that measures of experienced happiness would be more sensitive. I eventually assembled an interdisciplinary team to develop a measure of experienced happiness (Kahneman, Krueger, Schkade, Stone, and Schwarz, 2004), and we set out to demonstrate the aspiration treadmill. Over several years, we asked substantial samples of women to reconstruct a day of their life in detail. They indicated the feelings they had experienced during each episode, and we computed a measure of experienced happiness: the average quality of affective experience during the day. Our hypothesis was that differences in life circumstances would have more of an effect on this measure than on life satisfaction. We were so convinced that when we got our first batch of data, comparing teachers in top-rated schools to teachers in inferior schools, we actually misread the results as confirming our hypothesis. In fact, they showed the opposite: The groups of teachers differed more in their work satisfaction than in their affective experience at work. This was the first of many such findings: Income, marital status, and education all influence experienced happiness less than satisfaction, and we could show that the difference is not a statistical artifact. Measuring experienced happiness turned out to be interesting and useful, but not in the way we had expected. We had simply been wrong.

Experienced happiness, we learned, depends mainly on personality and on the hedonic value of the activities to which people allocate their time. Life circumstances influence the allocation of time, and the hedonic outcome is often mixed: High-income women have more enjoyable activities than the poor, but they also spend more time engaged in work that they do not enjoy; married women spend less time alone but more time doing tedious chores. Conditions that make people satisfied with their life do not necessarily make them happy.

Social scientists rarely change their minds, although they often adjust their position to accommodate inconvenient facts. But it is rare for a hypothesis to be so thoroughly falsified. Merely adjusting my position would not do; although I still find the idea of an aspiration treadmill attractive, I had to give it up.

To compound the irony, recent findings from the Gallup World Poll raise doubts about the puzzle itself. The most dramatic result is that when the entire range of human living standards is considered, the effects of income on a measure of life satisfaction (the "ladder of life") are not small at all. We had thought income effects are small because we were looking within countries. The GDP differences between countries are enormous and highly predictive of differences in life satisfaction. In a sample of over 130,000 people from 126 countries, the correlation between the life satisfaction of individuals and the GDP of the country in which they live was over .40—an exceptionally high value in social science. Humans everywhere, from Norway to Sierra Leone, apparently evaluate their life by a common standard of material prosperity, which changes as GDP increases. The implied conclusion—that citizens of different countries do not adapt to their level of prosperity—flies against everything we thought we knew ten years ago. We have been wrong and now we know it. I suppose this means that there is a science of well-being, even if we are not doing it very well.

Founder of The Whole Earth Catalog; cofounder of The Well and the Global Business Network; author of The Clock of the Long Now

Good Old Stuff Sucks

In the 1990s, I was praising the remarkable grassroots success of the building preservation movement. Keep the fabric and continuity of the old buildings and neighborhoods alive! Revive those sash windows.

As a landlocked youth in Illinois I mooned over the yacht sales pictures in the back of sailboat books. I knew what I wanted—a gaff-rigged ketch. Wood, of course.

The Christmas mail-order-catalog people know what my age group wants (I'm sixty-nine). We want to give a child wooden blocks, Monopoly or Clue, a Lionel train. We want to give ourselves a bomber jacket, a fancy leather belt, a fine cotton shirt. We study the Restoration Hardware catalog. My own *Whole Earth Catalog*, back when, pushed no end of retro stuff in a back-to-basics agenda.

Well, I bought a sequence of wooden sailboats. Their gaff rigs couldn't sail to windward. Their leaky wood hulls and decks were a maintenance nightmare. I learned that the fiberglass hulls we'd all sneered at were superior in every way to wood.

Remodeling an old farmhouse two years ago and replacing its sash windows, I discovered the current state of window technology. A standard Andersen window, factory-made exactly to the dimensions you want, has superb insulation qualities; superb hinges, crank, and lock; a flick-in/flick-out screen; and it looks great. The same goes for the new kinds of doors, kitchen cabinetry, and even furniture feet that are available—all drastically improved.

The message finally got through. Good old stuff sucks. Sticking with the fine old whatevers is like wearing 100 percent cotton in the mountains; it's just stupid.

Give me 100 percent not-cotton clothing, genetically modified food (from a farmers' market, preferably), this year's laptop, cutting-edge dentistry and drugs.

The Precautionary Principle tells me I should worry about everything new because it might have hidden dangers. The handwringers should worry more about the old stuff. It's mostly crap.

(New stuff is mostly crap, too, of course. But the best new stuff is invariably better than the best old stuff.)

Chief news and features editor of Nature; *author of* Mapping Mars

Against Human Spaceflight

I have, falteringly and with various intermediate about-faces and caveats, changed my mind about human spaceflight. I am of the generation to have had its childhood imagination stoked by the sight of Apollo missions on the television. I can't put hand on heart and say I remember the Eagle landing, but I remember the sights of the moon relayed to our home. I was fascinated by space and only through that, by way of the science fiction that a fascination with space inexorably led to, by science. And astronauts were what space was about.

I was not uncritical of human spaceflight as I grew older. I remember my anger at the *Challenger* explosion, my sense that if people were going to die, it should be for something grander than just another shuttle mission. But I was still struck by its romance and the way its romance touched some of the unlikeliest people. By all logic *The Economist* should have been, when I worked there, highly dubious about the aspirations of human spaceflight, as it is today. But the then-editor would hear not a word against the undertaking—at least, not against its principle.

With some relief at this, I became, as the magazine's science editor, a sort of critical apologist—critical of the human-space program but sensitive to the possibility that a better human-space program was possible.

I bought into, at some level, the argument that a joint U.S.-Russian program offered advantages in terms of aerospace employment in the former U.S.S.R. I bought into the argument that continuity of effort was needed—that so much would be lost if a program were dismantled that it might not be possible to reassemble it. I bought into the crucial safety-net argument— that it would not be possible to cancel the U.S. program anyway, so strong were the interests of the military-industrial complex and so broad (if shallow) the support of the public. (Like the Powder River, a mile wide, an inch deep, and rolling uphill all the way from Texas.) And I could see science it would offer that was unavailable by any other means.

Now, though, I can no longer find much to respect in those arguments. U.S.-Russian cooperation seems to have bought little benefit. The idea of continuous effort seems at best unproven— and, indeed, perhaps worth checking; leaving a technology fallow for a few decades and coming back with new people, tools, and mindsets is not such a bad idea. And at least one serious presidential candidate is talking about actually freezing the American program, canceling the shuttle without, in the short term, developing its successor. Whether Barack Obama will be elected or be willing or able to carry through this idea remains to be seen—but if politicians are talking like this, the "It will never happen, so why worry?" argument becomes far more suspect.

And the crucial idea (crucial to me) that human exploration of Mars might answer great questions about life in the universe no longer seems as plausible or as likely to pay off in my lifetime as once it did. I increasingly think that life in a Martian deep biosphere, if there is any, will be related to Earth life and teach us relatively little that's new. At the same time, it will be fiendishly hard to reach without contamination. Mars continues to fascinate me—but it has ever less need of a putative future human presence in order to do so.

My excitement at the idea of life in the universe—excitement undoubtedly spurred by Apollo and the works of Arthur Clarke, Robert Heinlein, and Gene Roddenberry that followed on from it in my education—is now more engaged with exoplanets, for which human spaceflight is entirely irrelevant (though post-human spaceflight may be a different kettle of lobsters). If we want to understand the depth of the various relations between life and planets, which is what I want to understand, it is by studying other planets with vibrant biospheres, as well as this one, that we will do so. A world with a spartan hundred-billion-dollar moon base but no ability to measure spectra and light curves from Earth-like planets around distant stars is not the world for me.

In general, I try to avoid arguing from my own interests. But in this case it seems to me that all the other arguments against human spaceflight are so strong that to be against it merely meant realizing that an atavistic part of me had failed to under-stand what those interests are. I'm interested in how life works on astronomical scales, and that interest has nothing to do, in the short term, with human spaceflight. And I see no reason beyond my own interests to suggest that it is something worth spending so much money on. It does not make the world a better place in any objective way that can be measured, or in any subjective way that compels respect.

It is possibly also the case that seeing human spaceflight reduced to a matter of suborbital hops for the rich, or even low-Earth-orbit hotels, has hardened my heart further against it. I hope this is not a manifestation of the politics of envy, though I fear that in part it could be.

JUDITH RICH HARRIS

Independent investigator and theoretician; author of No Two
Alike: Human Nature and Human Individuality

The Generalization Assumption

Anyone who has taken a course in introductory psychology has heard the story of how the behaviorist John B. Watson produced "conditioned fear" of a white rat—or was it a white rabbit?—in an unfortunate infant called Little Albert, and how Albert "generalized" that fear to other white, furry things (including, in some accounts, his mother's coat). It was a vividly convincing story, and like my fellow students I saw no reason to doubt it. Nor did I see any reason, until many years later, to read Watson's original account of the experiment, published in 1920. What a mess! You could find better methodology at a high school science fair. Not surprisingly—at least, it doesn't surprise me now—Watson's experiment has not stood up well to attempts to replicate it. But the failures to replicate are seldom mentioned in the introductory textbooks.

The idea of generalization is a very basic one in psychology. Psychologists of every stripe take it for granted that learned responses—behaviors, emotions, expectations, and so on—generalize readily and automatically to other stimuli of the same general type. It is assumed, for example, that once the baby has learned that his mother is dependable and his brother is aggressive, he will expect other adults to be dependable and other children to be aggressive.

I now believe that generalization is the exception, not the rule. Careful research has shown that babies arrive in the world with a bias against generalizing. This is true for learned motor skills and it is also true for expectations about people. Babies are born with the desire to learn about the beings who populate their world and the ability to store information about each individual separately. They do not expect all adults to behave like their mother or all children to behave like their siblings. Children who quarrel incessantly with their brothers and sisters usually get along much better with their peers. A firstborn who is accustomed to dominating his younger siblings at home is no more likely than a later-born to try to dominate his schoolmates on the playground. A boy's relationship with his father does not form the template for his later relationship with his boss.

I am not, of course, the only one in the world who has given up the belief in ubiquitous generalization, but if we formed a club we could probably hold meetings in my kitchen. Confirmation bias—the tendency to notice things that support one's assumptions and ignore or explain away anything that doesn't fit—keeps most people faithful to what they learned in intro psych. They observe that the child who is agreeable or timid or conscientious at home tends, to a certain extent, to behave in a similar manner outside the home, and they interpret this correlation as evidence that the child learns patterns of behavior at home which she then carries along with her to other situations.

The mistake they are making is to ignore the effects of genes. Studies using advanced methods of data analysis have shown that the similarities in behavior from one context to another are due chiefly to genetic influences. Our inborn predispositions to behave in certain ways go with us wherever we go, but learned behaviors are tailored to the situation. The fact that genetic predispositions tend to show up early is the reason why some psychologists also make the mistake of attributing too much importance to early experiences.

What changed my mind about these things was the realization that if I tossed out the assumption about generalization, some hitherto puzzling findings about human behavior suddenly made more sense. I was fifty-six years old at the time but fairly new to the field of child development, and I had no stake in maintaining the status quo. It is a luxury to have the freedom to change one's mind.

Computational neuroscientist, Salk Institute for Biological Studies; coauthor (with Patricia Churchland) of The Computational Brain

Spike Timing in Cortical Neurons

How is it that insects manage to get by on many fewer neurons than we have? A fly brain has a few hundred thousand neurons, compared with the few hundred billion in our brains—a million times more neurons. Flies are quite successful in their niche. They can see, find food, mate, and create the next generation of flies. The traditional view is that unique neurons evolved in the brain of the fly to perform specific tasks, in contrast to the mammalian strategy of creating many more neurons of the same type, working together in a collective fashion. This view was bolstered when it became possible to record from single cortical neurons, which responded to sensory stimuli with highly variable spike trains from trial to trial. Reliability could be achieved only by averaging the responses of many neurons.

Theoretical analysis of neural signals in large networks assumed statistical randomness in the responses of neurons. These theories used the average firing rates of neurons as the primary statistical variable. Individual spikes and the times when they occurred were not relevant in these theories. In contrast, the timing of single spikes in flies has been shown to carry specific information about sensory stimuli important for guiding the behavior of flies, and in mammals the timing of spikes in the peripheral auditory system carries information about the spatial locations of sound sources. However, cortical neurons did not seem to care about the timing of spikes.

I have changed my mind about cortical neurons and now think they are far more capable than we ever imagined. Two important experimental results pointed me in this direction. First, if you repeatedly inject the same fluctuating current into a neuron in a cortical slice, to mimic the inputs that occur in an intact piece of tissue, the spike times are highly reproducible from trial to trial. This shows that cortical neurons are capable of initiating spikes with millisecond precision. Second, if you arrange for a single synapse to be stimulated a few milliseconds just before or just after a spike in the neuron, the synaptic strength will increase or decrease, respectively. This tells us that the machinery in the cortex is every bit as capable as a fly brain, but what is it being used for?

The cerebral cortex is constantly being bombarded by sensory inputs and has to sort though the myriad of signals for those that are the most important and respond selectively to them. The cortex also needs to organize the signals being generated internally, in the absence of sensory inputs. The hypothesis I have been pursuing over the last decade is that spike timing in cortical neurons is used internally as a way of controlling the flow of communication between neurons. This is different from the traditional view that spike times code sensory information, as occurs in the periphery. Rather, spike timing and the synchronous firing of large numbers of cortical neurons may be used to enhance the salience of sensory inputs, as occurs during focal attention, and to decide what information is worth saving for future use. According to this view, the firing rates of neurons are used as an internal representation of the world, but the timing of spikes is used to regulate the communication of signals between cortical areas.

The way neuroscientists perform experiments is biased by their theoretical views. If cortical neurons use rate coding, you need to record, and report, only their average firing rates. But to find out if spike timing is important, new experiments need to be designed and new types of analysis need to be performed on the data. Neuroscientists have begun to pursue these new experiments, and we should know before too long where they will lead us.

JONATHAN HAIDT

Psychologist, University of Virginia; author of The Happiness Hypothesis

Hanging Out with the Boys

I was born without the neural cluster that makes boys find pleasure in moving balls and pucks around through space and talking endlessly about men who get paid to do such things. I always knew I could never join a fraternity or the military, because I wouldn't be able to fake the sports talk. By the time I became a professor, I had developed the contempt widespread in academe for any institution that brings young men together to do groupish things. Primitive tribalism, I thought. Initiation rites, alcohol, sports, sexism, and baseball caps turn decent boys into knuckleheads. I'd have gladly voted to ban fraternities, ROTC, and most sports teams from my university.

But not anymore. Three books convinced me I had misunderstood such institutions because I had too individualistic a view of human nature. The first book was David Sloan Wilson's *Darwin's Cathedral*, which argued that human beings were shaped by natural selection operating simultaneously at multiple levels, including the group level. Humans went through a major transition in evolution when we developed religiously inclined minds and religious institutions that activated those minds, binding people into groups capable of extraordinary cooperation without kinship.

The second book was William McNeill's *Keeping Together in Time*, about the historical prevalence and cultural importance of synchronized dance, marching, and other forms of movement. McNeill argued that such "muscular bonding" was an evolutionary innovation, an "indefinitely expansible basis for social cohesion among any and every group that keeps together in time."

The third book was Barbara Ehrenreich's *Dancing in the Streets*, which made the same argument as McNeill but with much more attention to recent history and to the concept of *communitas* or group love. Most traditional societies had group dance rituals that functioned to soften structure and hierarchy and to increase trust, love, and cohesion. Westerners too have a need for *communitas*, Ehrenreich argues, but our society makes it hard to satisfy it, and our social scientists have little to say about it.

These three books gave me a new outlook on human nature. I began to see us not just as chimpanzees with symbolic lives but also as bees without hives. When we made the transition over the last two hundred years from tight communities (*Gemeinschaft*) to free and mobile societies (*Gesellschaft*), we escaped from bonds that were sometimes oppressive, yes, but into a world so free it left many of us gasping for connection, purpose, and meaning. I began to think about the many ways people, particularly young people, have found to combat this isolation. Rave parties and the Burning Man festival are spectacular examples of new ways to satisfy the ancient longing for *communitas*. But suddenly sports teams, fraternities, and even the military made a lot more sense.

I now believe that such groups do great things for their members and often create social capital and other benefits that spread beyond their borders. The strong school spirit and alumni loyalty we all benefit from at the University of Virginia would drop sharply if fraternities and major sports were eliminated. If my son grows up to be a sports-playing fraternity brother, a part of me may still be disappointed. But I'll give him my blessing, along with three great books to read.

Professor of ethology, Cambridge University; author of Design for a Life: How Biology and Psychology Shape Human Behavior

Standing Up for Atheism

Near the end of his life, Charles Darwin invited for lunch at Down House Dr. Ludwig Büchner, president of the Congress of the International Federation of Freethinkers, and Edward Aveling, a self-proclaimed and active atheist. The invitation was at their request. Emma Darwin, devout as ever, was appalled by the thought of entertaining such guests and at table insulated herself from the atheists with an old family friend, the Reverend Brodie Innes, on her right and her grandson and his friends on her left. After lunch Darwin and his son Frank smoked cigarettes with the two visitors in Darwin's old study. Darwin asked his guests with surprising directness, "Why do you call yourselves atheists?" Darwin said he preferred the word "agnostic." While he agreed that Christianity was not supported by evidence, he felt that "atheist" was too aggressive a term to describe his own position.

For many years, what had been good enough for Darwin was good enough for me. I, too, described myself as an agnostic.

I had been brought up in a Christian culture, and some of the most rational humanists I knew were believers. I loved the music and art that had been inspired by a belief in God and saw no hypocrisy in participating in the great carol services held in the chapel of King's College Cambridge. I did not accept the views of some of my scientific colleagues that the march of science had disposed of religion. The wish that I and many biologists had to understand biological evolution was not the same as the wish had by those with deep religious conviction to understand the meaning of life.

I had, however, led a sheltered existence and had never met anybody who was aggressively religious. I hated what I read about the ugly fanaticism of all forms of religious fundamentalism—or what I saw of it on television. However, such wickedness did not seem to be simply correlated with religious belief, since many nonbelievers were just as totalitarian in their behavior as the believers.

My unwillingness to be involved in religious debates was shaken at a grand dinner party. The woman sitting next to me asked me what I did, and I told her that I was a biologist. "Oh well," she said, "then we have plenty to talk about, because I believe that every word of the Bible is literally true." My heart sank.

As things turned out, we didn't have a great deal to talk about, because she wasn't going to be persuaded by any argument that I could throw at her. She did not seem to wonder about the inconsistencies between the gospels of the New Testament or between the first and second chapters of Genesis. Nor was she concerned about where Cain's wife came from. The Victorians were delicate about such matters and were not going to entertain the thought that Cain married an unnamed sister, or, horrors, that his own mother bore his children, his grandchildren, and so on down the line of descendants until other women became available. Nevertheless, the devout Victorians were obviously troubled by the question, and they speculated on the existence of pre-Adamite people—angels, probably—who would have furnished Cain with his wife.

My creationist dinner companion was not worried by such trivialities and dismissed my lack of politesse as the problem of a scientist being too literal. However, being too literal was not my problem, it was hers and those of her fellow creationists. She was hoist with her own petard. In any event, it was quite simply stupid to try to take on science on its own terms by appealing to the intelligence implicit in natural design. Science provides orderly methods for examining the natural world. One of those methods is to develop theories that integrate as much as possible of what we know about the phenomena encompassed by the theory. The theories provide frameworks for testing the characteristics of the world—and though some theorists may not wish to believe it, their theories are eminently disposable. Facts are widely shared opinions and every so often the consensus breaks—and minds change. Nevertheless it is crying for the moon to hope that the enormous bodies of thought that have been built up about cosmology, geology, and biological evolution are all due to fall apart. No serious theologian would rest his or her beliefs on such a hope. If faith rests on the supposed implausibility of a current scientific explanation, it is vulnerable to the appearance of a plausible one. To build on such sand is a crass mistake.

Not long after that dreadful dinner, Richard Dawkins wrote to me to ask whether I would publicly affirm my atheism. I could see no reason why not. One of the clear definitions of an atheist is a lack of a belief in a God. That certainly described my position, even though I am disinclined to attack the beliefs of the sincere and thoughtful people with strong religious beliefs whom I continue to meet. I completed the questionnaire that Richard had sent to me. I had changed my mind. A dear friend,

Peter Lipton, who died suddenly in November 2007, had been assiduous in maintaining Jewish customs in his own home and in his public defense of Israel. After he died, I was surprised to discover that he described himself as a religious atheist. I should not have been surprised.

ALAN ALDA

Actor, writer, and director; host of PBS's Scientific American Frontiers

Without a God

So far, I've changed my mind twice about God.

Until I was twenty, I was sure there was a being who could see everything I did and who didn't like most of it. He seemed to care about minute aspects of my life, like on what day of the week I ate a piece of meat. And yet he let earthquakes and mudslides take out whole communities, apparently ignoring the saints among them who ate their meat on the assigned days. Eventually, I realized I didn't believe there was such a being. It didn't seem reasonable. And I assumed I was an atheist.

As I understood the word, it meant that I was someone who didn't believe in a God; I was without a God. I didn't broadcast this in public, because I noticed that people who do believe in a god get upset to hear that others don't. (Why this is so is one of the most pressing of human questions, and I wish a few of the bright people in this conversation would try to answer it through research.)

But slowly I realized that in the popular mind the word "atheist" was coming to mean something more: a statement that there *couldn't* be a God. God was, in this formulation, not possible, and this was something that could be proved. But I had been changed by eleven years of interviewing six or seven hundred scientists around the world on the television program *Scientific American Frontiers*. And that change was reflected in how I would now identify myself.

The most striking thing about the scientists I met was their complete dedication to evidence. It reminded me of the wonderfully plainspoken words of Richard Feynman, who felt it was better not to know than to believe something that was wrong. The problem for me was that just as I couldn't find any evidence that there was a god, I couldn't find any that there wasn't a god. I would have to call myself an agnostic. At first, this seemed a little wimpy, but after a while I began to hope it might be an example of Feynman's heroic willingness to accept, even glory in, uncertainty.

I still don't like the word "agnostic." It's too fancy. I'm simply not a believer. But as simple as this notion is, it confuses some people. Someone wrote a Wikipedia entry about me, identifying me as an atheist because I'd said in a book I wrote that I wasn't a believer. I guess in a world uncomfortable with uncertainty, an unbeliever must be an atheist, and possibly an infidel. This gets us back to that most pressing of human questions: Why do people worry so much about other people's holding beliefs other than their own? This is the question that makes the subject over which I changed my mind something of global importance and not just a personal, semantic dalliance.

Do our beliefs identify us, the way our language, foods, and customs do? Is this why people who think the universe chugs along on its own are as repellent to some as people who eat live monkey brains are to others? Are we saying, "You threaten my identity with your infidelity to my beliefs—you're trying to kill me with your thoughts, so I'll get you first with this stone"? And, if so, is this really something that can be resolved through reasonable discourse?

Maybe this is an even more difficult problem, one that's written in the letters that spell out our DNA. Why is the belief in God and gods so ubiquitous? Does belief in a higher power confer some slight health benefit, and has natural selection favored those who are genetically inclined to believe in such a power — and is that why so many of us are inclined to believe? (Whether or not a god actually exists, the tendency to believe we'll be saved might give us the strength to escape sickness and disaster and live the extra few minutes it takes to replicate ourselves.)

These are wild speculations, of course, and they're probably based on a desperate belief I once had that we could one day understand ourselves.

But I might have changed my mind on that one, too.

Psychologist, Harvard University; author of The Stuff of Thought

Have Humans Stopped Evolving?

Ten years ago, I wrote:

> *For ninety-nine percent of human existence, people lived as foragers in small nomadic bands. Our brains are adapted to that long-vanished way of life, not to brand-new agricultural and industrial civilizations. They are not wired to cope with anonymous crowds, schooling, written language, government, police, courts, armies, modern medicine, formal social institutions, high technology, and other newcomers to the human experience.*

And:

> *Are we still evolving? Biologically, probably not much. Evolution has no momentum, so we will not turn into the creepy bloat-heads of science fiction. The modern human condition is not conducive to real evolution either. We infest the whole habitable and not-so-habitable earth,*

*migrate at will, and zigzag from lifestyle to lifestyle. This
makes us a nebulous, moving target for natural selection.
If the species is evolving at all, it is happening too slowly
and unpredictably for us to know the direction.*

(*from* How the Mind Works)

Though I stand by a lot of those statements, I've had to question the overall assumption that human evolution pretty much stopped by the time of the agricultural revolution. When I wrote those passages, completion of the Human Genome Project was several years away, and so was the use of statistical techniques that test for signs of selection in the genome. Some of these searches for "Darwin's fingerprint," as the technique has been called, have confirmed predictions I had made. For example, the modern version of the gene associated with language and speech has been under selection for several hundred thousand years and has even been extracted from a Neanderthal bone, consistent with my hypothesis (with Paul Bloom) that language is a product of gradual natural selection. But the assumption of no-recent-human-evolution has not.

New results from the labs of Jonathan Pritchard, Robert Moyzis, Pardis Sabeti, and others have suggested that thousands of genes, perhaps as much as 10 percent of the human genome, have been under strong recent selection, and the selection may even have accelerated during the past several thousand years. The numbers are comparable to those for maize, which has been artificially selected beyond recognition during the past few millennia.

If these results hold up and apply to psychologically relevant brain function (as opposed to disease resistance, skin color, and digestion, which we already know have evolved in recent millennia), then the field of evolutionary psychology might have to reconsider the simplifying assumption that biological evolution was pretty much over and done with ten thousand to fifty thousand years ago.

And if so, the result could be evolutionary psychology on steroids. Humans might have evolutionary adaptations not just to the conditions that prevailed for hundreds of thousands of years but also to some of the conditions that have prevailed only for millennia or even centuries. Currently, evolutionary psychology assumes that any adaptations to post-agricultural ways of life are 100 percent cultural.

Though I suspect some revisions will be called for, I doubt they will be radical, for two reasons. One is that many aspects of the human (and ape) environments have been constant for a much longer time than the period in which selection has recently been claimed to operate. Examples include dangerous animals and insects, toxins and pathogens in spoiled food and other animal products, dependent children, sexual dimorphism, risks of cuckoldry and desertion, parent-offspring conflict, risk of cheaters in cooperation, fitness variation among potential mates, causal laws governing solid bodies, presence of conspecifics with minds, and many others. Recent adaptations would have to be an icing on this cake—quantitative variations within complex emotional and cognitive systems.

The other is the empirical fact that human races and ethnic groups are psychologically highly similar, if not identical. People everywhere use language, get jealous, are selective in choosing mates, find their children cute, are afraid of heights and the dark, experience anger and disgust, learn names for local species, and so on. If you adopt children from a technologically undeveloped part of the world, they will fit into modern society just fine. To the extent that this is true, there can't have been a whole lot of uneven psychological evolution postdating the split among the races fifty thousand to a hundred thousand years ago (though there could have been parallel evolution in all the branches).

NICHOLAS A. CHRISTAKIS

Physician and social scientist, Harvard University; author of
Death Foretold: Prophecy and Prognosis in Medical Care

Evolution in Real Time

I work in a borderland between social science and medicine, and I therefore often find myself trying to reconcile conflicting facts and perspectives about human biology and behavior. There are fellow travelers at this border, of course, heading in both directions, or just dawdling, but the border is both sparsely populated and chaotic. The border is also, strangely, well patrolled, and it is often quite hard to get authorities on both sides to coordinate activities. Once in a while, however, I find that my passport (never quite in order, according to officials) has acquired a new visa. This past year, I acquired the conviction that human evolution may proceed much faster than I had thought, and that humans themselves may be responsible.

In short, I have changed my mind about how people come to embody the social world around them. I once thought we internalized cultural factors by forming memories, acquiring language, or bearing emotional and physical marks (of poverty, of conquest). I thought this was the limit of the ways in which our bodies were shaped by our social environment. In particular, I thought our genes were historically immutable and that it was not possible to imagine a conversation between culture and genetics. I thought we as a species evolved over time frames far too long to be influenced by human actions.

I now think this is wrong, and that the alternative—that we are evolving in real time, under the pressure of discernible social and historical forces—is true. Rather than a monologue of genetics or a soliloquy of culture, there is a dialectic between genetics and culture.

Evidence has been mounting for a decade. The best example so far is the evolution of lactose tolerance in adults. The ability of adults to digest lactose (a sugar in milk) confers evolutionary advantages only when a stable supply of milk is available, such as after milk-producing animals (sheep, cattle, goats) have been domesticated. The advantages are several, ranging from a source of valuable calories to a source of necessary hydration during times of water shortage or spoilage. Amazingly, just over the last three thousand to nine thousand years there have been several adaptive mutations in widely separated populations in Africa and Europe, all conferring the ability to digest lactose (as shown by Sarah Tishkoff and others). These mutations are principally seen in populations that are herders, and not in nearby populations that have retained a hunter-gatherer lifestyle. This trait is sufficiently advantageous that those with the trait have many more descendants than those without.

A similar story can be told about mutations that have arisen in the relatively recent historical past that confer ability to survive epidemic diseases, such as typhoid. Since these diseases were made more likely when the density of human settlements increased and far-flung trade became possible, here we have another example of how culture may affect our genes.

But this past year, a paper by John Hawks and colleagues in the *Proceedings of the National Academy of Sciences* functioned like the staccato *plunk* of a customs agent stamping my documents and waving me on. The paper showed that the human genome may have been changing at an accelerating rate over the past eighty thousand years, and that this change may be in response not only to population growth and adaptation to new environments but also to cultural developments that have made it possible for humans to sustain such population growth or survive in such environments.

Our biology and our culture have always been in conversation of course, just not (I had thought) on the genetic level. For example, rising socioeconomic status with industrial development results in people becoming taller (a biological effect of a cultural development) and taller people require changes in architecture (a cultural effect of a biological development). Anyone who has ever marveled at the small size of beds in colonial-era houses knows this firsthand. Similarly, an epidemic may induce large-scale social changes, modifying kinship systems or political power. But genetic change over short time periods? Yes.

Why does this matter? Because it is hard to know where this would stop. There may be genetic variants that favor survival in cities, that favor saving for retirement, that favor consumption of alcohol, that favor a preference for complicated social networks. There may be genetic variants (based on altruistic genes that are a part of our hominid heritage) that favor living in a democratic society, others that favor living among computers, still others that favor certain kinds of visual perception (maybe we are all more myopic as a result of Medieval lens grinders). Modern cultural forms may favor some traits over others. Maybe even the more complex world we live in nowadays really is making us smarter.

This has been very difficult for me to accept, because, unfortunately, this also means that it may be the case that particular ways of living create advantages for some but not all members of our species. Certain groups may acquire (admittedly, over centuries) certain advantages, and there might be positive or negative feedback loops between genetics and culture. Maybe some of us really are better able to cope with modernity than others. The idea that what we choose to do with our world modifies what kind of offspring we have is as amazing as it is troubling.

PAUL DAVIES

Physicist, Arizona State University; author of The Goldilocks
Enigma: Why Is the Universe Just Right for Life?

Laws as Emergent with the Universe

I used to be a committed Platonist.

For most of my career, I believed that the bedrock of physical reality lay with the laws of physics — magnificent, immutable,
transcendent, universal, infinitely precise mathematical relationships that rule the universe with as sure a hand as that of any
god. And I had orthodoxy on my side, for most of my physicist
colleagues also believe that these perfect laws are the levitating
superturtle that holds up the mighty edifice we call nature, as
disclosed through science. About three years ago, however, it
dawned on me that such laws are an extraordinary and unjusti-
fied idealization.

How can we be sure that the laws are infinitely precise? How
do we know they are immutable and apply without the slightest
change from the beginning to the end of time? Furthermore,
the laws themselves remain unexplained. Where do they come
from? Why do they have the form that they do? Indeed, why do
they exist at all? And if there are many possible such laws, then,
as Stephen Hawking has expressed it, what is it that "breathes

fire" into a particular set of laws and makes a universe for them to govern?

So I did a U-turn and embraced the notion of laws as emergent with the universe rather than stamped on it from without like a maker's mark. The "inherent" laws I now espouse are not absolute and perfect but intrinsically fuzzy and flexible, although for almost all practical purposes we don't notice the tiny flaws.

Why did I change my mind? I am not content to merely accept the laws of physics as a brute fact; rather, I want to explain the laws, or at least explain the form they have, as part of the scientific enterprise. One of the oddities about the laws is the well-known fact that they are weirdly well suited to the emergence of life in the universe. Had they been slightly different, chances are there would be no sentient beings around to discover them.

The fashionable explanation for this—that there is a multiplicity of laws in a multiplicity of parallel universes, with each set of laws fixed and perfect within its host universe—is a nice try but still leaves a lot unexplained. And simply saying that the laws "just are" seems no better than declaring, "God made them that way."

The orthodox view of perfect physical laws is a thinly veiled vestige of monotheism, the reigning worldview that prevailed at the birth of modern science. If we want to explain the laws, however, we have to abandon the theological legacy that the laws are fixed and absolute and replace them with the notion that the states of the world and the laws that link them form a dynamic interdependent unity.

LEO M. CHALUPA

Neurobiologist, University of California, Davis

Constancy of the Persona

The hottest topic in neuroscience today is brain plasticity. This catchphrase refers to the fact that various types of experience can significantly modify key attributes of the brain. This field began decades ago by focusing on how different aspects of the developing brain could be affected by early rearing conditions.

More recently, the field of brain plasticity has shifted to studies demonstrating a remarkable degree of change in the connections and functional properties of mature and even aged brains. Thousands of published papers have now appeared on this topic, many by reputable scientists, and this has led to a host of books, programs, and even commercial enterprises touting the malleability of the brain with "proper training." One is practically made to feel guilty for not taking advantage of this thriving store of information to improve one's own brain or those of one's children and grandchildren.

My field of research is developmental neurobiology, and I used to be a proponent of the potential benefits documented by brain plasticity studies. I am still of the opinion that brain plasticity is a real phenomenon, one that deserves further study and one

231

that could be utilized to better human welfare. But my careful reading of this literature has tempered my initial enthusiasm.

For one thing, those selling a commercial product are making many of the major claims for the benefits of brain-exercise regimes. It is also the case that my experiences outside the laboratory have caused me to question the limitless potential of brain plasticity advocated by some devotees.

Point of fact: Recently I had the chance to meet someone I had not seen since childhood. The person had changed physically beyond all recognition, as might be expected. Yet after spending some time with this individual, his personality traits of long ago became apparent, including a rather peculiar laugh I remember from grade school.

Point of fact: A close colleague had a near-fatal car accident, one that kept him in a coma for many days and in intensive care for weeks thereafter. Shortly after returning from his ordeal, this A-type personality changed into a seemingly mellow and serene person. But in less than two months, even before the physical scars of his accident had healed, he was back to his old driven self.

For a working scientist to invoke anecdotal experience to question a scientific field of endeavor is akin to heresy. But it seems to me that it is foolish to simply ignore what one has learned from a lifetime of experiences. The older I get, the more my personal interactions convince me that a person's core remains remarkably stable in spite of huge experiential variations. With all the recent emphasis on brain plasticity, there has been virtually no attempt to explain the stability of the individual's core attributes, values, and beliefs.

Here is a real puzzle to ponder: Every cell in your body, including all one hundred billion neurons in your brain, is in a constant process of breakdown and renewal. Your brain is different from the one you had a year or even a month ago, even without special brain exercises. So how is the constancy of one's persona maintained? The answer to that question offers a far greater challenge to our understanding of the brain than the currently voguish field of brain plasticity.

Anthropologist, Centre National de la Recherche Scientifique, France; author of In Gods We Trust: The Evolutionary Landscape of Religion

Friendship and Faith

I am an anthropologist who has traveled to many places and met many different kinds of people. I try to know what it is like to be someone very different from me in order to better understand what it means to be human. But it is only in the last few years that my thinking has deeply changed on what drives major differences between animal and human behavior, such as willingness to kill and die for a cause.

I once thought that the "Paleolithic Revolution" in human culture that led to the expansion of our species across the planet beginning about 50,000 years ago was triggered by a cognitive mutation, most probably for language. But the latest DNA evidence indicates humans emerged in Eastern and Southern Africa 200,000 years ago, and just stayed there for another 150,000 years while Neanderthals migrated from Western Europe to Siberia. The human population may have been on the verge of extinction, dwindling to as few as two thousand souls 70,000 years ago. Then, in a geological instant, one or a few hardy human bands burst out of Africa, first to Australia, then the Middle East and Central Asia, then Europe and the Americas.

So here's the puzzle: Assuming humans had pretty much the same anatomy and cognitive capacity since they first punctuated the African scrub, how come they did nothing much that was culturally human for most of human existence?

My guess is that for much of that time there was little competition between human bands and that the critical novelty was a pressing need to cooperate in order to compete. Thus friendship and teamwork with non-kin developed, which mitigated selfishness and made cultural life possible between genetic strangers. Think about how teams of friends bond in trust, rapidly acquire new information from one another, and immediately understand how to arrange themselves so as to best respond to a challenge in a life-threatening instance (a battle, a hunt, an impending boat wreck). And faith in that friendship, in something bigger and more lasting than the morning mist, seems to me what all religious and political life is about.

Here's an anecdote that kick-started me thinking about this.

While preparing a psychological experiment on limits of rational choice with Muslim mujahedin on the Indonesian island of Sulawesi, I noticed tears welling in my traveling companion and bodyguard, Farhin (who had earlier hosted 9/11 mastermind Khalid Sheikh Mohammed in Jakarta and helped to blow up the Philippines' ambassador's residence). Farhin had just heard of a young man recently killed in a skirmish with Christian fighters.

"Farhin," I asked, "you knew the boy?"

"No," he said, "but he was only in the jihad a few weeks. I've been fighting since Afghanistan and am still not a martyr."

I tried consoling with my own disbelief: "But you love your wife and children."

"Yes." He nodded sadly. "God has given this, and I must have faith in His way."

I had come to the limits of my understanding of the other. There was something in Farhin that was incalculably different from me, yet almost everything else was not.

"Farhin, in all those years, after you and the others came back from Afghanistan, how did you stay a part of the jihad?" I asked.

I expected him to tell me about his religious fervor and devotion to a Great Cause.

"The [Indonesian] Afghan Alumni never stopped playing soccer together," he replied matter-of-factly. "That's when we were closest together in the camp." He smiled. "Except when we went on vacation to fight the communists, we played soccer and remained brothers."

Maybe people don't kill and die simply for a cause. They do it for friends—campmates, schoolmates, workmates, soccer buddies, body-building buddies, pin-ball buddies—who share a cause. Some die for dreams of jihad—of justice and glory—but nearly all in devotion to a family-like group of friends and mentors, of "fictive kin."

Then it became embarrassingly obvious: It is no accident that nearly all religious and political movements express allegiance through the idiom of the family—Brothers and Sisters, Children of God, Fatherland, Motherland, Homeland, and the like. Nearly all such movements require subordination, or at least assimilation, of any real family (genetic kinship) to the larger imagined community of "Brothers and Sisters." Indeed, the complete subordination of biological loyalty to ideological loyalty for the Ikhwan, the "Brotherhood" of the Prophet, is Islam's original meaning: "Submission."

My research team has analyzed every attack by Farhin and his friends, who belong to Southeast Asia's Jemmah Islamiyah (JI). I have interviewed key JI operatives (including cofounder Abu Bakr Ba'asyir) and counterterrorism officials who track JI.

Our data show that support for suicide actions is triggered by moral outrage at perceived attacks against Islam and sacred values, but this is converted to action as a result of small-world factors. Out of millions who express sympathy with global jihad, only a few thousand show willingness to commit violence. They tend to go to violence in small groups, consisting mostly of friends and some kin. These groups arise within specific "scenes": neighborhoods, schools (classes, dorms), workplaces, and common leisure activities (soccer, mosque, barbershop, café, online chat rooms).

Three other examples:

1. In Al Qaeda, about 70 percent join with friends, 20 percent with kin. Our interviews with friends of the 9/11 suicide pilots reveal they weren't "recruited" into Qaeda. They were Middle Eastern Arabs isolated in a Moroccan Islamic community in a Hamburg suburb. Seeking friendship, they started hanging out after mosque services in local restaurants and barbershops, eventually living together when they self-radicalized. They wanted to go to Chechnya, then Kosovo, only landing in a Qaeda camp in Afghanistan as a distant third choice.

2. Five of the seven plotters in the 2004 Madrid train bombing who blew themselves up when cornered by police grew up in the tumble-down neighborhood of Jemaa Mezuaq in Tetuan, Morocco. In 2006, at least five more young Mezuaq men went to Iraq on "martyrdom missions." One in the Madrid group was related to one in the Iraq group by marriage; each group included a pair of brothers. All went to the same elementary school, all but one to the same high school. They played soccer as friends, went to the same mosque, mingled in the same cafés.

3. Hamas's most sustained suicide bombing campaign in 2003–2004 involved several buddies from Hebron's Masjad (mosque) al-Jihad soccer team. Most lived in the Wad Abu Katila neighborhood and belonged to the al-Qawasmeh hamula (clan); several were classmates in the neighborhood's local branch of the Palestine Polytechnic College.

Social psychology tends to support the finding that group-think often trumps individual volition and knowledge, whether in our society or any other. But for Americans bred on a constant diet of individualism, the group is not where one generally looks for explanation. This was particularly true for me, but the data caused me to change my mind.

MARCO IACOBONI

Neuroscientist, Brain Mapping Center, University of California, Los Angeles; author of Mirroring People: The New Science of How We Connect with Others

The Marginal Role of Science

Some time ago, I thought rational, enlightened thinking would eventually eradicate irrational thinking and supernatural beliefs. How could it be otherwise? Scientists and enlightened people have facts and logical arguments on their side, whereas people "on the other side" have only unprovable beliefs and bad reasoning. I was wrong, way wrong. Thirty years later, irrational thinking and supernatural beliefs are much stronger than they used to be, permeate ours and other societies, and do not seem to be going away anytime soon. How is it possible? Shouldn't history always move forward? What went wrong? What can we do to fix this backward movement toward the irrational?

The problem is that science still plays a marginal role in our public discourse. Indeed, there are no science books on the *New York Times* 100 Notable Books of the Year list, there is no science category in *The Economist's* Books of the Year 2007, and only Oliver Sacks appears on *The New Yorker's* list of Books From Our Pages.

Why does science play this marginal role? I think there's more than one reason. First, scientists tend to confine themselves in well-defined, narrow boundaries. They tend not to claim any wisdom outside the confines of their specialties. By doing so, they marginalize themselves and make it difficult for science to have an effect on society. It is high time for scientists to step up and claim wisdom outside their specialty.

There are other ways, however, to have an effect on society—for instance, by making changes in scientific practice. These days, scientific practice is dominated by the hypothesis-testing paradigm. While there is nothing wrong with hypothesis testing, it is definitely wrong to confine all science to it. This approach precludes the study of complex real-world phenomena, the phenomena important to people outside academia. It is time to perform more broad-based descriptive studies on issues relevant to our society.

Another dominant practice in science (definitely in neuroscience, my field) is to study phenomena from an atemporal perspective. Only the timeless seems to matter to most neuroscientists. Even time itself tends to be studied from this "Platonic ideal" perspective. I guess this approach stems from the general tendency of science to adopt the detached "view from nowhere," as the philosopher Thomas Nagel puts it. If we have learned anything from modern science, however, it is that there is no such thing, no view from nowhere. It is time for scientists, especially neuroscientists, to commit to the study of the finite and temporal. The issues that matter here and now are the issues that people relate to.

How should we do this? One way of disseminating the scientific method in our public discourse is to use the tools and approaches of science to investigate issues salient to the general public. In neuroscience, we now have powerful tools that let us do this. We can study how people make decisions and form affiliations—not from a timeless perspective but from the perspective of "here and now." These are the kinds of studies that naturally engage people. Reading about such studies, people are more likely to learn scientific facts (even the atemporal ones) and absorb the scientific method and reasoning. My hope is that by being exposed to and engaged by scientific facts, methods, and reasoning, people will eventually find it difficult to believe unprovable things.

RICHARD WRANGHAM

*Professor of biology and anthropology, Harvard University;
coauthor (with Dale Peterson) of* Demonic Males: Apes and
the Origins of Human Violence

Hominoid Cuisine

Like people since even before Darwin, I used to think human origins were explained by meat-eating. But three epiphanies have changed my mind. I now think cooking was the major advance that made us human.

First, an improved fossil record has shown that meat-eating arose too early to explain human origins. Significant meat-eating by our ancestors is initially attested in the prehuman world of 2.6 million years ago, when hominids began to flake stones into simple knives. Around the same time, there appears a fossil species variously called *Australopithecus habilis* or *Homo habilis*. These *habilis* presumably made the stone knives, but they were not human. They were Calibans, missing links with an intricate mixture of advanced and primitive traits. Their brains, being twice the size of ape brains, tell of incipient humanity, but as Bernard Wood has stressed, their chimpanzee-sized bodies, long arms, big guts, and jutting faces made them apelike. Meat-eating likely explains the origin of *habilis*.

Humans emerged almost a million years later, when *habilis* evolved into *Homo erectus*. At 1.6 million years ago, *Homo erectus* were the size and shape of people today. Their brains were bigger than those of *habilis*, and they walked and ran as easily as we do. Their mouths were small and their teeth relatively dwarfed—a pygmy-faced hominoid, just like all later humans. To judge from the reduced flaring of their rib cages, they had lost the capacious guts that allow great apes and *habilis* to eat large volumes of plant food. Equally strange for a supposed helpless and defenseless species, they had also lost their climbing ability, forcing them to sleep on the ground—a surprising commitment in a continent full of big cats, sabre-tooths, hyenas, rhinos, and elephants.

So the question of what made us human is the question of why a population of *habilis* became *Homo erectus*. My second epiphany was a double insight: Humans are biologically adapted to eating cooked diets, and the signs of this adaptation start with *Homo erectus*. Cooked food is the signature feature of human diet. It not only makes our food safe and easy to eat but it also grants us large amounts of energy, compared with a raw diet, obviating the need to ingest big meals. Cooking softens food, too, thereby making eating so speedy that as eaters of cooked food we are granted many extra hours of free time every day.

So cooked food allows our guts, teeth, and mouths to be small, while giving us abundant food energy and freeing our time. Cooked food, of course, requires the control of fire, and a fire at night explains how *Homo erectus* dared sleep on the ground.

Cooked food has so many important biological effects that its adoption should be clearly marked in the fossil record by signals of a reduced digestive system and increased energy use. While such signs are clear at the origin of *Homo erectus*, they are not found later in human evolution. The match between the biological merits of cooked food and the evolutionary changes in *Homo erectus* is thus so obvious that except for a scientific obstacle, I believe it would have been noticed long ago. The obstacle is the insistence of archaeologists that the control of fire is not firmly evidenced before about a quarter of a million years ago. As a result of this archaeological caution, the idea that humans could have used fire before about 250,000 to 500,000 years ago has long been sidelined.

But I finally realized that the archaeological record decays so steadily that it gives us no information about when fire was first controlled. The fire record is better at 10,000 years than at 20,000 years, at 50,000 years than at 100,000 years, at 250,000 years than at 500,000 years, and so on. Evidence for the control of fire is always better when it is closer to the present, but in the course of human evolution it never completely goes away. There is only one date beyond which no evidence for the control of fire has been found: 1.6 million years ago, around the time when *Homo erectus* evolved. Between now and then, the erratic record tells us only one thing: The archaeological evidence is incapable of telling us when fire was first controlled. The biological evidence is more helpful. That was my third epiphany.

The origin of *Homo erectus* is too late for meat-eating; the adoption of cooking solves the problem; and archaeology does not gainsay it. In a roast potato and a hunk of beef we have a new theory of what made us human.

SEAN CARROLL

Theoretical physicist, California Institute of Technology; author of Spacetime and Geometry: An Introduction to General Relativity

How Not to Overthrow the System

Growing up as a young proto-scientist, I was always strongly antiestablishmentarian, looking forward to overthrowing the System as our generation's new Galileo. Now I spend a substantial fraction of my time explaining and defending the status quo to outsiders. It's very depressing.

As an undergraduate astronomer, I was involved in a novel and exciting test of Einstein's general relativity: measuring the precession of orbits, just like Mercury in the solar system but using massive eclipsing binary stars. What made it truly exciting was that the data disagreed with the theory. (Which they still do, by the way.) How thrilling is it to have the chance to overthrow Einstein himself? Of course, there are more mundane explanations—the stars are tilted, or there is an invisible companion star perturbing their orbits, and these hypotheses were duly considered. But I wasn't very patient with such boring possibilities. It was obvious to me that we had dealt a crushing blow to a cornerstone of modern physics and the Establishment was just too hidebound to admit it.

Now I know better. Physicists who are experts in the field tend to be skeptical of experimental claims that contradict general relativity, not because they are hopelessly encumbered by tradition but because Einstein's theory has passed a startlingly diverse array of experimental tests. Indeed, it turns out to be almost impossible to change general relativity in a way that would be important for those binary stars but which would not have already shown up in the solar system. Experiments and theories don't exist in isolation; they form a tightly connected web in which changes to any one piece tend to reverberate through various others.

So now I find myself cast as a defender of scientific orthodoxy—from classics like relativity and natural selection to modern wrinkles like dark matter and dark energy. In science, no orthodoxy is sacred or above question; there should always be a healthy exploration of alternatives, and I have always enjoyed inventing new theories of gravity or cosmology, keeping in mind the variety of evidence in favor of the standard picture. But there is also an unhealthy brand of skepticism proceeding from ignorance rather than expertise, which insists that any consensus must flow from a reluctance to face up to the truth rather than an appreciation of the evidence. It's that kind of skepticism that keeps showing up in my e-mail. Unsolicited.

Heresy is more romantic than orthodoxy. Nobody roots for Goliath, as Wilt Chamberlain was fond of saying. But in science ideas tend to grow into orthodoxy for good reasons: They fit the data better than the alternatives. Many casual heretics can't be bothered with all the detailed theoretical arguments and experimental tests supporting the models they hope to overthrow. They have a feeling about how the universe should work, and they're convinced that history will eventually vindicate them, just as it did Galileo.

What they fail to appreciate is that, scientifically speaking, Galileo overthrew the system from within. He understood the reigning orthodoxy of his time better than anyone, so he was better able to see beyond it. Our present theories are not complete, and nobody believes they are the final word on how nature works. But finding the precise way to make progress, to pinpoint the subtle shift of perspective that will illuminate a new way of looking at the world, will require an intimate familiarity with our current ideas, and a respectful appreciation of the evidence supporting them.

Being a heretic can be fun, but being a successful heretic is mostly hard work.

LINDA STONE

Cofounder and former director of Microsoft's Virtual Worlds Group/Social Computing Group

Pay Attention!

In the past few years, I have been thinking and writing about "attention," and specifically "continuous partial attention." The impetus came from my years of working at Apple, and then, Microsoft, where I thought a lot about user interface as well as our relationship to the tools we create.

I believe that attention is the most powerful tool of the human spirit and that we can enhance or augment our attention with practices like meditation and exercise, diffuse it with technologies like e-mail and BlackBerries, or alter it with pharmaceuticals.

But lately I have observed that the way in which many of us interact with our personal technologies makes it impossible to use this extraordinary tool of attention to our advantage.

In observing others—in their offices, their homes, at cafés—I have noticed that most people hold their breath when they begin responding to e-mail. On cell phones, especially when talking and walking, people tend to hyperventilate. Both of these breathing patterns disturb oxygen and CO_2 balance.

Research conducted by two NIH scientists, Margaret Chesney and David Anderson, demonstrates that breath-holding can contribute significantly to stress-related diseases. The body becomes acidic, the kidneys begin to reabsorb sodium, and as the oxygen and CO_2 balance is undermined, our biochemistry is thrown off.

Around this time, I became interested in the vagus nerve and the role it plays. The vagus nerve is one of the major cranial nerves, and wanders from the head to the neck, chest, and abdomen. Its primary job is to mediate the autonomic nervous system, which includes the sympathetic ("fight or flight") and parasympathetic ("rest and digest") nervous systems.

The parasympathetic nervous system governs our sense of hunger and satiety, flow of saliva and digestive enzymes, the relaxation response, and many aspects of healthy organ function. Focusing on diaphragmatic breathing enables us to downregulate the sympathetic nervous system, which then causes the parasympathetic nervous system to become dominant. Shallow breathing, breath-holding, and hyperventilating trigger the sympathetic nervous system, in a "fight or flight" response.

The activated sympathetic nervous system causes the liver to dump glucose and cholesterol into our blood, our heart rate increases, we don't have a sense of satiety, and our bodies anticipate and resource for the physical activity that historically accompanies a physical fight-or-flight response. When the only physical activity is sitting and responding to e-mail, we're sort of "all dressed up and no place to go."

Some breathing patterns favor our body's move toward parasympathetic functions and other breathing patterns favor a sympathetic nervous system response. Buteyko (breathing techniques developed by a Russian MD), Andy Weil's breathing exercises, diaphragmatic breathing, certain yoga breathing techniques —all have the potential to soothe us and help our bodies determine when fight or flight is really necessary and when we can rest and digest.

I've changed my mind about how much attention to pay to my breathing patterns and how important it is to remember to breathe when I'm using a computer, PDA, or cell phone. I've discovered that the more consistently I tune in to healthy breathing patterns, the clearer it is to me whether I'm hungry or not, the more easily I fall asleep and rest peacefully at night, and the more my outlook is consistently positive. I've come to believe that within the next five to seven years, breathing exercises will be a significant part of any fitness regime.

STANISLAS DEHAENE

Cognitive neuropsychology researcher, Institut National de la Santé, France; author of The Number Sense: How the Mind Creates Mathematics

The Brain's Equation

What made me change my mind wasn't a new fact but a new theory.

Although much of my work is dedicated to modeling the brain, I always thought that this enterprise would remain rather limited in scope. Unlike physics, neuroscience would never create a single, major, simple yet encompassing theory of how the brain works. There would never be a single "Schrödinger equation for the brain."

The vast majority of neuroscientists, I believe, share this pessimistic view. The reason is simple: The brain is the outcome of five hundred million years of tinkering. It consists of millions of distinct pieces, each evolved to solve a distinct yet important problem for our survival. Its overall properties result from an unlikely combination of thousands of receptor types, ad hoc molecular mechanisms, a great variety of categories of neurons, and, above all, a million billion connections criss-crossing the white matter in all directions. How could such a jumble be captured by a single mathematical law?

Well, I wouldn't claim that anyone has achieved that yet, but I have changed my mind about the possibility that such a law might exist.

For many theoretical neuroscientists, it all started twenty-five years ago when John Hopfield made us realize that a network of neurons could operate as an attractor network, driven to optimize an overall energy function that could be designed to accomplish object recognition or memory completion. Then came Geoff Hinton's Boltzmann machine: Again, the brain was seen as an optimizing machine that could solve complex probabilistic inferences. Yet both proposals were frameworks rather than laws. Each individual network realization still required the setup of thousands of ad hoc connection weights.

Very recently, however, Karl Friston, of University College London, has presented two extraordinarily ambitious and demanding papers in which he presents a "theory of cortical responses." Friston's theory rests on a single amazingly compact premise: The brain optimizes a free energy function. This function measures how closely the brain's internal representation of the world approximates the true state of the real world. From this simple postulate, Friston spins off an enormous variety of predictions: the multiple layers of cortex, the hierarchical organization of cortical areas, their reciprocal connection with distinct feed-forward and feedback properties, the existence of adaptation and repetition suppression, even the type of learning rule—Hebb's rule, or the more sophisticated spike-timing dependent plasticity—can be deduced, no longer postulated, from this single overarching law.

The theory fits easily within what has become a major area of research—the Bayesian brain, or the extent to which brains perform optimal inferences and take optimal decisions based on the rules of probabilistic logic. Alex Pouget, for instance, recently showed how neurons might encode probability distributions of parameters of the outside world, a mechanism that could be usefully harnessed by Fristonian optimization. And the physiologist Mike Shadlen has discovered that some neurons closely approximate the log-likelihood ratio in favor of a motor decision, a key element of Bayesian decision making. My colleagues and I have shown that the resulting random-walk decision process nicely accounts for the duration of a central decision stage, present in all human cognitive tasks, which might correspond to the slow, serial phase in which we consciously commit to a single decision. During nonconscious processing, my proposal is that we also perform Bayesian accumulation of evidence but without attaining the final commitment stage. Thus Bayesian theory is bringing us increasingly closer to the holy grail of neuroscience—a theory of consciousness.

Another reason I am excited about Friston's law is, paradoxically, that it isn't simple. It seems to have just the right level of distance from the raw facts. Much like Schrödinger's equation cannot easily be turned into specific predictions, even for an object as simple as a single hydrogen atom, Friston's theory requires heavy mathematical derivations before it ultimately provides useful outcomes. Not that it is inapplicable. On the contrary, it readily applies to motion perception, audio-visual integration, mirror neurons, and thousands of other domains—but in each case a rather involved calculation is needed.

It will take us years to decide whether Friston's theory is the true inheritor of Helmholtz's view of "perception as inference." What is certain, however, is that neuroscience now has a wealth of beautiful theories that should attract the attention of top-notch mathematicians—we will need them!

Cultural anthropologist; president of the Institute for Intercultural Studies; author of Willing to Learn: Passages of Personal Discovery

Making and Changing Minds

We do not so much change our mind about facts, although we necessarily correct and rearrange them in changing contexts. But we do change our minds about the significance of those facts.

I can remember, as a young woman, first grasping the danger of environmental destruction at a conference in 1968. The context was the intricate interconnection within all living systems, a concept that applied to ecosystems like forests and tide pools and equally well to human communities and the planet as a whole, the sense of an extraordinary interweaving of life, beautiful and fragile, and threatened by human hubris. It was at that conference that I first heard of the greenhouse effect, the mechanism that underlies global warming.

A few years later, I heard of the Gaia hypothesis, put forth by James Lovelock in 1970, which proposed that the same systemic interconnectivity gives the planet its resilience and a capacity for self-correction that might survive human tampering. Some environmentalists welcomed the Gaia hypothesis, while others warned that it might lead to complacency in the face of real and present danger. With each passing year, our knowledge of how things are connected is enriched, but the significance of these observations is still debated.

J. B. S. Haldane was asked once what the natural world suggested about the mind of its Creator, and he replied, "An inordinate fondness for beetles." This observation also plays differently for different listeners—a delight in diversity, perhaps, as if the Creator might have spent the first Sabbath afternoon resting from his work by playfully exploring the possible ramifications of a single idea (beetles make up roughly one-fifth of all known species on the planet, some 350,000 of them), or a humbling (or humiliating?) lack of preoccupation with our own unique kind, which might prove to be a temporary afterthought, survived only by cockroaches.

These two ways of looking at what we observe seem to recur, like the glass half full and the glass half empty. The more we know of the detail of living systems, the more we seem torn between anxiety and denial on the one hand and wonder and delight on the other as we try to understand the significance of our knowledge. Science has radically altered our awareness of the scale and age of the universe, but this changing awareness seems to stimulate humility in some—our planet a tiny speck dominated by flealike bipeds—and a sort of megalomania in others, who see all this as directed toward us, our species, as predestined masters. Similarly, the exploration of human diversity in the twentieth century expanded for some the sense of plasticity and variability and for others reinforced the sense of human unity. Even within these divergent emphases, for some the recognition of human unity includes a capacity for mutual recognition and adaptation while for others it suggests innate tendencies toward violence and xenophobia. As we have slowly explored the mechanisms of cultural change and adaptation, we have seen examples of (fragile) human communities demoralized by exposure to other cultures and (resilient) examples of extraordinary adaptability. At one moment, humans are depicted as potential stewards of the biosphere, at another as a cancer or a dangerous infestation. The growing awareness of a shared and interconnected destiny has a shadow side, the version of globalization that looks primarily for profit.

We are having much the same sort of debate at present between those who see religion primarily as a source of conflict between groups and others who see the world's religions as potentially convergent systems of meaning that have knit peoples together and laid the groundwork for contemporary ideas of human rights and civil society. Some believers feel called to treasure and respect the creation, including the many human cultures that have grown within it, while others regard differences of belief as sinful and the world we know as transitory or illusory. Each of the great religions, with different language and different emphases, offers the basis for environmental responsibility and for peaceful coexistence and compassion, but believers differ in what they choose to emphasize, all too many choosing the apocalyptic over the ethical texts. Nevertheless, major shifts have been occurring in the interpretation of information about climate change, most recently within the evangelical Christian community.

My guess is that many people have tilted first one way and then the other over the past fifty years, as we have become increasingly aware of diverse understandings—surprised by accounts of human creativity and adaptation and distressed at the resurgence of ancient quarrels and loss of tolerance and mutual respect. Some people are growing away from irresponsible consumerism, while others are having their first taste of affluence. Responses are probably partly based on temperament—generalized optimism versus pessimism—so the tension will not be resolved by scientific findings. But these responses are also based on the decisions we make, on making up our minds about which interpretations we choose to believe. The world's historical religions deal in different ways with limitation and with the need for sacrifice, but the materials are there for working together, just as they are there for stoking conflict and competition. We need to make up our minds. We are most likely to survive this century if we decide to approach the choices and potential losses ahead with an awareness of the risks we face but at the same time with an awareness of the natural wonders around us and a determination to deal with each other with respect and faith in the possibility of cooperation and responsibility.

CAROLYN PORCO

Planetary scientist; Cassini imaging science team leader; director of CICLOPS (Cassini Imaging Central Laboratory for OPerationS), Boulder, Colorado; adjunct professor, University of Colorado

The Practice of Science
Finds Itself Imperiled

I've changed my mind about the manner in which our future on this planet might evolve.

I used to think that the power of science to dissect, inform, illuminate, and clarify; its venerable record in improving the human condition; and its role in enabling the technological progress of the modern world were all so glaringly obvious that no one could reasonably question its hallowed position in human culture as the preeminent device for separating truth from falsehood.

I used to think that the edifice of knowledge constructed from thousands of years of scientific thought by various cultures all over the globe, and in particular the insights earned over the last four hundred years from modern scientific methods, were so universally revered that we could feel comfortably assured of having permanently left our philistine days behind us.

And while I've always appreciated the need for care and perseverance in guiding public evaluation of the complexities of scientific discourse and its findings, I never expected we would, at this stage in our development, have to justify and defend the scientific process itself.

Yet that appears to be the case today. Now I'm no longer sure that scientific inquiry and the value it places on verifiable truth can survive without constant protection, and its ebb and flow over the course of human history affirms this. We have been beset in the past by dark ages, when scientific truths and the ideas that logically spring from them were systematically destroyed or made otherwise unavailable, when the practitioners of science were discredited, imprisoned, and even murdered. Periods of human enlightenment have been the exception throughout time, not the rule, and even our language acknowledges this. "Two steps forward, one step back" neatly encapsulates the non-monotonic stagger inherent in any reading of human history.

And, if we're not mindful, we could stagger again. When the truth becomes problematic, when intellectual honesty clashes with political expediency, when voices of reason are silenced to mere whispers, when fear alloys with ignorance to promote physical might over intelligence, integrity, and wisdom, the practice of science can find itself imperiled. At that point, can darkness be far behind?

To avoid so dangerous a tipping point requires us to recognize the distasteful possibility that it could happen again at any time. I now suspect the danger will be forever present, the need for vigilance forever great.

AUBREY DE GREY

Gerontologist; chairman and chief science officer of the Methuselah Foundation; author of Ending Aging

Science and Technology: Joined at the Hip?

The words "science" and "technology," or equivalently the words "research" and "development," are used in the same breath so readily that one might easily presume they are joined at the hip, that their goals are indistinguishable, and that those who are good at one are, if not necessarily equally good at the other, at least quite good at evaluating the quality of work in the other. I grew up with this assumption, but the longer I work at the interface between science and technology, the more I find myself having to accept that it is false—that most scientists are rather poor at the type of thinking that identifies efficient new ways to get things done and, likewise, that most technologists are not terribly good at identifying efficient ways to find things out.

I've come to feel that there are several reasons underlying this divide. A major one is the divergent approaches of scientists and technologists to the use of evidence. In basic research, it is exceptionally easy to be seduced by one's data—to see a natural interpretation of it and to overlook the existence of other, comparably economical interpretations of it that lead to dramatically different conclusions. It therefore makes sense for scientists to give the greatest weight, when evaluating the evidence for and against a given hypothesis, to the most direct observational or experimental evidence at hand.

Technologists, however, succeed best when they stand back from the task before them, thinking laterally about ways in which ostensibly irrelevant techniques might be applied to solve one or another component of the problem. The technologist's approach, when applied to science, is likely to result all too often in wasted time, as experiments are performed that contain too many departures from previous work to allow the drawing of firm conclusions either way concerning the hypothesis of interest.

Conversely, applying the scientist's methodology to technological endeavors can also result in wasted time, resulting from overly small steps away from techniques already known to be futile, like trying to fly by flapping mechanical wings.

But there's another difference between the characteristic mindsets of scientists and technologists, and I view it as the most problematic. Scientists are avowedly curiosity-driven rather than goal-directed—they are spurred by the knowledge that, throughout the history of civilization, innumerable useful technologies have become possible not through the stepwise execution of a predefined plan but rather through the purposely undirected quest for knowledge, letting a dynamically determined sequence of experiments lead where it may.

That logic is as true as it ever was, and any technologist who doubts it need only examine the recent history of science. However, it can be—and, in my view, all too often is—taken too far. A curiosity-driven sequence of experiments is useful not because of the sequence but because of the technological opportunities that emerge at the end of the sequence. The sequence is not an end in itself. And this is rather important to keep in mind. Any scientist, on completing an experiment, is spoiled for choice concerning what experiment to do next—or, more prosaically, concerning what experiment to apply for funding to do next.

The natural criterion for making this choice is the likelihood that the experiment will generate a wide range of answers to technologically important questions, thereby providing new technological opportunities. But an altogether more frequently adopted criterion, in practice, is that the experiment will generate a wide range of new questions—new reasons to do more experiments. This is only indirectly useful, and in practice it is indeed less frequently useful than programs of research designed with one eye on the potential for eventual technological utility.

Why, then, is it the norm? Simply because it is the more attractive to those who are making these decisions—the curiosity-driven scientists (whether grant applicants or grant reviewers) themselves. Curiosity is addictive; both emotionally and in their own enlightened self-interest, scientists want reasons to do more science, not more technology. But as a society we need science to be as useful as possible, as quickly as possible, and this addiction slows us down.

HELENA CRONIN

Philosopher, London School of Economics; director and founder of Darwin@LSE; author of The Ant and the Peacock: Altruism and Sexual Selection from Darwin to Today

"More Dumbbells but More Nobels"

What gives rise to the most salient, contested, and misunderstood of sex differences—differences that see men persistently walk off with the top positions and prizes, whether in influence or income, whether as heads of state or CEOs . . . differences that infuriate feminists, preoccupy policymakers, galvanize legislators, and spawn "diversity" committees and degrees in gender studies?

I used to think that these patterns of sex differences resulted mainly from average differences between men and women in innate talents, tastes, and temperaments. After all, men are on average more mathematical and more technically minded, women more verbal; men are more interested in things, women are more interested in people; men are more competitive, risk-taking, single-minded, status-conscious, women far less so. Therefore, even where such differences are modest, the distribution of these three Ts among males will necessarily be different from that among females—and so will give rise to notable differences between the two groups. Add to this some bias and barriers—a sexist attitude here, lack of child care there. And the sex differences are explained. Or so I thought.

But I have now changed my mind. Talents, tastes, and temperaments play fundamental roles, but they don't fully explain the differences. It is a fourth T that most decisively shapes the distinctive structure of male-female differences. That T is *tails*, the tails of these statistical distributions. Females are much of a muchness, clustering round the mean. But among males the variance—the difference between the most and the least, the best and the worst—can be vast. So males are almost bound to be overrepresented both at the bottom and at the top. I think of this as "More dumbbells but more Nobels."

Consider the mathematics sections in the National Academy of Sciences: 95 percent male. Which contributes most to this predominance—higher means or larger variance? One calculation yields the following answer. If the sex difference between the means were obliterated but the variance were left intact, male membership would drop modestly to 91 percent. But if the means were left intact but the difference in the variance were obliterated, male membership would plummet to 64 percent. The overwhelming male predominance stems largely from greater variance.

Similarly, consider the most intellectually gifted of the U.S. population, an elite 1 percent. The difference between their bottom and top quartiles is so wide that it encompasses one-third of the entire ability range in the American population, from IQs above 137 to IQs beyond 200. And who's overwhelmingly in the top quartile? Males. Look, for instance, at the boy:girl ratios among adolescents for scores in mathematical-reasoning tests: scores of at least 500, 2:1; scores of at least 600, 4:1; scores of at least 700, 13:1.

Admittedly, those examples are writ large—exceptionally high aptitude and a talent that strongly favors males and with a notably long right-hand tail. Nevertheless, the same combined causes—the forces of natural selection and the facts of statistical distribution—ensure that this is the default template for male-female differences.

Let's look at those causes. The legacy of natural selection is twofold: mean differences in the three Ts and males generally being more variable. These two features hold for most sex differences in our species, and, as Darwin noted, greater male variance is ubiquitous across the entire animal kingdom. As to the facts of statistical distribution, they are threefold (and watch what happens at the end of the right-hand tail): First, for overlapping bell curves, even with only a small difference in the means, the ratios become more inflated as one goes farther out along the tail; second, where there's greater variance, there's likely to be a dumbbells-and-Nobels effect; and third, when one group has both greater mean and greater variance, that group becomes even more overrepresented at the far end of the right tail.

The upshot? When we're dealing with evolved sex differences, we should expect that the farther out we go along the right curve, the more we will find men predominating. So there we are: Whether or not there are more male dumbbells, there will certainly be—both figuratively and actually—more male Nobels.

Unfortunately, however, this is not the prevailing perspective in current debates, particularly where policy is concerned. On the contrary, discussions standardly zoom in on the means and blithely ignore the tails. So sex differences are judged to be small. And thus it seems that there's a gaping discrepancy: If women are as good on average as men, why are men overwhelmingly at the top? The answer must be systematic unfairness— bias and barriers. Therefore, so the argument runs, it is to bias and barriers that policy should be directed. And so the results of straightforward facts of statistical distribution get treated as political problems—as "evidence" of bias and barriers that keep women back and sweep men to the top. (Though how this explains the men at the bottom is an unacknowledged mystery.)

But science has given us biological insights, statistical rules, and empirical findings. Surely, if the bias-and-barriers believers care more about science and facts than stance and faith, they too should be changing their minds about men at the top.

Physicist, Massachusetts Institute of Technology; recipient of the 2004 Nobel Prize in Physics; author of Lightness of Being: Mass, Ether, and the Unification of Forces

Against Debunking

I was an earnest student in catechism class. The climax of our early training, as thirteen-year-olds, was an intense retreat in preparation for the sacrament of confirmation. Even now I vividly remember the rapture of belief, the glow that everyday events acquired when I felt they reflected a grand scheme of the universe, in which I had a personal place. Soon afterward, though, came disillusionment. As I learned more about science, some of the concepts and explanations in the ancient sacred texts came to seem clearly wrong, and as I learned more about history and how it is recorded, some of the stories in those texts came to seem very doubtful.

What I found most disillusioning, however, was not that the sacred texts contained errors but that they suffered by comparison. Compared with what I was learning in science, they offered few truly surprising and powerful insights. Where was there a vision to match the concepts of infinite space, of vast expanses of time, of distant stars that rivaled and surpassed our sun? Or of hidden forces and new, invisible forms of "light"? Or of tremendous energies that humans could, by understanding natural processes, learn to liberate and control? I came to think that if God exists, He (or She, or They, or It) did a much more impressive job revealing Himself in the world than in the old books, and that the power of faith and prayer is elusive and unreliable compared with the daily miracles of medicine and technology.

For many years, like some of my colleagues and some recent bestselling authors, I thought that active, aggressive debunking might be in order. I've changed my mind. One factor was my study of intellectual history. Many of my greatest heroes in physics, including Galileo, Newton, Faraday, Maxwell, and Planck, were deeply religious people. They truly believed that what they were doing in their scientific studies was discovering the mind of God. Many of Bach's and Mozart's most awesome productions are religiously inspired. Saint Augustine's writings display one of the most impressive intellects ever. And so on. Can you imagine hectoring this group? And what would be the point? Did their religious beliefs make them stupid, or stifle their creativity?

Also, debunking hasn't worked very well. David Hume set out the main arguments for religious skepticism in the early eighteenth century. Bertrand Russell and many others have augmented them since. Textual criticism reduces fundamentalism to absurdity. Modern molecular biology, rooted in physics and chemistry, demonstrates that life is a natural process; Darwinian evolution illuminates its natural origin. These insights have been highly publicized for many decades, yet religious doctrines that contradict some or all of them have not merely survived but prospered.

Why? Part of the answer is social. People tend to stay with the religion of their birth, for the same sorts of reasons that they stay loyal to their clan, or their country.

But beyond that, religion addresses some deep concerns that science does not yet, for most people, touch. The human yearning for meaningful understanding, our fear of death—these deep motivations are not going to vanish.

Understanding, of course, is what science is all about. Many people imagine, however, that scientific understanding is dry and mundane, with no scope for wonder and amazement. That is simply ignorant. Looking for wonder and amazement? Try some quantum theory!

Beyond understanding interconnected facts, people want to discover their significance or meaning. Neuroscientists are beginning to map human motivations and drives at the molecular level. As this work advances, we will attain a deeper understanding of the meaning of meaning. Freud's theories had enormous impact not because they are correct but because they "explained" why people feel and act as they do. Correct and powerful theories that address these issues are sure to have much greater influence.

Meanwhile, medical science is taking a deep look at aging. Within the next century, it may be possible for people to prolong youth and good health for many years—perhaps indefinitely. This would, of course, profoundly change our relationship with death. So to me the important challenge is not to debunk religion but to address its issues in better ways.

Editor in chief of Nature

Self-Improvement

I've changed my mind about the use of enhancement drugs by healthy people. A year ago, I'd have been against the idea, whereas now I think there's much to be said for it.

The ultimate test of such a change of mind is how I'd feel if my offspring (both adults) went down that road, and my answer is that with tolerable risks of side effects and zero risk of addiction, then I'd feel OK if there was an appropriate purpose to it. "Appropriate purposes" exclude gaining an unfair advantage or unwillingly following the demands of others but include gaining a better return on an investment of study or of developing a skill.

I became interested in the issues surrounding cognitive enhancement as one example of debates about human enhancement—debates that can only get more vigorous in future. It's also an example of a topic in which both natural and social sciences can contribute to better regulation, another theme that interests me. Thinking about the issues and looking at the evidence-based literature made me realize how shallow was my own instinctive aversion to the use of such drugs by healthy people. It also led to a thoughtful article by Barbara Sahakian and Sharon Morein-

Zamir in *Nature* (20 December 2007) that triggered many blog discussions.

Social scientists report that a small but significant proportion of students on at least some campuses are using prescription drugs in order to help their studies—drugs such as modafinil (prescribed for narcolepsy) and methylphenidate (prescribed for attention-deficit/hyperactivity disorder). I've not seen studies that quantify similar use by academic faculty, or by people in other nonmilitary walks of life, though there is no doubt that it is happening. There are anecdotal accounts and experimental small-scale trials showing that such drugs do indeed improve performance to a modest degree under particular circumstances.

New cognitive-enhancing drugs are being developed, officially for therapy. And the therapeutic importance—both current and potential—of such drugs is indeed significant. But manufacturers won't turn away the significant revenues from illegal use by the healthy.

That word "illegal" is the rub. Off-prescription use is illegal—in the United States, at least. But that illegality reflects an official drug culture that is highly questionable. It's a culture in which the Food and Drug Administration seems reluctant generally to embrace the regulation of enhancement for the healthy, though empowered to do so. It is also a culture that is rightly concerned about risk but wrongly founded in the idea that drug use by healthy people is by definition a Bad Thing. That in turn reflects instinctive attitudes having to do with "naturalness" and "cheating yourself" that don't stand up to rational consideration. Perhaps more to the point, they don't stand up to behavioral considerations, as Viagra has shown.

Research and societal discussions are necessary before cognitive-enhancement drugs should be made legally available for the healthy, but I now believe that that is the right direction in which to head.

DANIEL GOLEMAN

Psychologist; author of Social Intelligence

The Inexplicable Monks

The sociologist Anselm Strauss was a proponent of methods to generate "grounded theory," that is, a progressive series of hypotheses that are tested, then refined according to what the data shows, and tested again, and so refined, in a perpetual cascade of theory-data loops, each of which presents new conclusions and raises new questions. In this model, the essence of the scientific method boils down to changing your mind for the right reasons, and asking the right questions.

And now it's happened to me; I've changed my mind yet again.

One of my most basic assumptions about the relationship between mental effort and brain function has begun to crumble. Here's why.

My earliest research interests as a psychologist were in the ways mental training can shape biological systems. My doctoral dissertation was a psychophysiological study of meditation as an intervention in stress reactivity; I found (as have many others since) that the practice of meditation seems to speed the rate of physiological recovery from a stressor.

My guiding assumptions included the standard premise that the mind-body relationship operates according to orderly, understandable principles. One such might be called the "dose-response" rule that the more time put into a given method of training, the greater the result in the targeted biological system. This is a basic correlate of neuroplasticity, the mechanism through which repeated experience shapes the brain.

For example, a string of research has now established that more experienced meditators recover more quickly from stress-induced physiological arousal than do novices. Nothing remarkable there; the dose-response rule would predict this is so. Thus brain-imaging studies show that the spatial areas of London taxi drivers become enhanced during the first six months they spend driving around that city's winding streets; likewise, the area for thumb movement in the motor cortex becomes more robust in violinists as they continue to practice over many months.

This relationship has been confirmed in many varieties of mental training. A seminal 2004 article in the *Proceedings of the National Academy of Sciences* found that compared with novices, highly adept meditators generated far more high-amplitude gamma-wave activity—which reflects finely focused attention—in areas of the prefrontal cortex while meditating.

The seasoned meditators in this study—all Tibetan lamas—had undergone cumulative levels of mental training akin to the amount of lifetime sports practice put in by Olympic athletes: ten thousand to fifty thousand hours. Novices tended to increase gamma activity by around 10 to 15 percent in the key brain area, while most experts had increases on the order of 100 percent from baseline. What caught my eye in this data was not this difference between novices and experts (which might be explained in any number of ways, including a self-selection bias) but rather a discrepancy in the data among the group of Olympic-level meditators.

Although the experts' average boost in gamma was around 100 percent, two lamas were "outliers": their gamma levels leapt 600 to 800 percent. This goes far beyond an orderly dose-response relationship—these jumps in high-amplitude gamma activity are the highest ever reported in the scientific literature apart from pathological conditions like seizures. Yet the lamas were voluntarily inducing this extraordinarily heightened brain activity for just a few minutes at a time—and by meditating on "pure compassion," no less.

I have no explanation for this data, but plenty of questions. At the higher reaches of contemplative expertise, do principles apply (as the Dalai Lama has suggested in dialogues with neuroscientists) that we do not yet grasp? If so, what might these be? In truth, I have no idea. But these puzzling data points have pried open my mind a bit as I've had to question what had been a rock-solid assumption of my own.

Or so I thought. All the above was what I wrote for the annual *Edge* Question 2008: "What have you changed your mind about? Why?"

A few weeks later I happened to be talking about this answer with Richard Davidson, a neuroscientist at the University of Wisconsin, and one of the authors of the study with the remarkable meditators. He pointed out to me that the findings for the two lamas were not statistical outliers, but fit the regression analysis as shown by a scatter plot tucked away among the article's data tables. These two lamas were the champions among the Olympic-level meditators, having put in the highest number of lifetime retreat hours—about 44,000 and 55,000 hours—and also showing the greatest effect from all that mind training. That's exactly what the dose-response model predicts.

So now I've changed my mind a second time. I no longer see these data points as inexplicable in terms of neuroplasticity. Now I see them as a first scientific report from the upper reaches of neural transformation. So my change of mind has to do with what might be possible at those upper reaches of human consciousness.

I'm left with a new set of questions. At the highest reaches of mind training, I wonder, do a novel range of possibilities for self-regulating biological functions emerge? What could be the actual experience of these intense amplifications of neural activity? And since this remarkable brain activity occurred during the cultivation of compassion, what might the human benefits be of such training?

DAVID M. BUSS

Psychologist, University of Texas at Austin; author of The Murderer Next Door: Why the Mind Is Designed to Kill

The Sexual Strategies of Women

I have never thought that female sexual psychology was simple, but I've changed my mind about the magnitude of its complexity and consequently revamped the scope and orchestration of my entire research program. I once focused my research on two primary sexual strategies—long-term and short-term. Empirical work has revealed a deeper, richer repertoire: serial mating, friends with benefits, one-night stands, brief affairs, enduring affairs, polyamory, polyandry, sexual mate poaching, mate expulsion, mate switching, and various combinations of these throughout life. Women implement their sexual strategies through an astonishing array of tactics. Scientists have documented at least thirty-four distinct tactics for promoting short-term sexual encounters and nearly double that for attracting a long-term romantic partner.

Researchers discovered twenty-eight tactics women use to derogate sexual competitors, from pointing out that her rival's thighs are heavy to telling others that the rival has a sexually transmitted disease. Women's sexual strategies include at least nineteen tactics of mate retention, ranging from vigilance to violence, and twenty-nine tactics of ridding themselves of unwanted mates, including having sex as a way to say good-bye. Some women use sexual infidelity as a means of getting benefits from two or more men. Others use it as a means of exiting one relationship in order to enter another. When a woman wants a man who is already in a relationship, she can use at least nineteen tactics of mate poaching to lure him away, from befriending both members of the couple in order to disarm her unsuspecting rival to insidiously sowing seeds of doubt about her rival's fidelity or level of desirability.

Ovulation and orgasm are yielding scientific insights into female sexuality unimagined five years ago. The hidden rhythms of the ovulation cycle, for example, have profound effects on women's sexual desire. Women married to men lower in mate-value experience an upsurge in sexual fantasies about other men, but mainly during the fertile phase of their cycle. They are sexually attracted to men with masculine faces, but especially so in the five days leading up to ovulation. Women's sense of smell spikes around ovulation. Sexual scents, long thought unimportant in human sexuality, in fact convey information to women about a man's genetic quality. The female orgasm, once thought by many scientists to be functionless, may turn out to have several distinct adaptive benefits. And those don't even include the potential gains from faking orgasm. Some women mislead about their sexual satisfaction in order to get a man to leave; others to deceive him about his paternity.

Female sexual psychology touches every facet of human affairs, from cooperative alliances through strategies of hierarchy negotiation. Some women use sex to get along. Some use sex to get ahead. Sexual motives pervade murder. Failure in sexual unions sometimes triggers suicidal ideation. I thought the complexity of women's sexual psychology was finally starting to be

278 -- DAVID M. BUSS

captured when recent research revealed 237 reasons why women have sex, ranging from "to get rid of a headache" to "to get closer to God," from "to become emotionally connected with my partner" to "to break up a rival's relationship." Within a month of that publication, however, researchers discovered 44 more reasons why women have sex, ranging from "because life is short and we could die at any moment" to "to get my boyfriend to shut up," bringing the sexual motivation total to 281 and counting. (Obviously, trying to pin down exact numbers is a bit of a joke, but scientists work through quantification.)

Yet with all these scientific discoveries, I feel we are still at the beginning of the exploration and I am humbled by how little we still know. As a researcher focusing on female sexuality, I'm inherently limited, by virtue of possessing a male brain. Consequently, I've teamed up with brilliant female research scientists, recruited a team of talented female graduate students, and marshaled much of my research to explore the complexities of female sexual psychology. These steps have led me to see things previously invisible to my male-blinkered brain. Female sexual psychology is more complex than I previously thought by several orders of magnitude. And still I may be underestimating.

ROBERT SHAPIRO

Chemist, New York University; author of Planetary Dreams: The Quest to Discover Life Beyond Earth

Smothering Science with Silence

I used to view the scientific literature as a collective human effort to build an enduring and expanding structure of knowledge. Each new publication in a respected, refereed journal would be digested and debated with the thoroughness that religious groups devote to the Talmud, Bible, or Koran. In science, of course, new papers can challenge widely held beliefs, so publication does not mean acceptance. The alternative is criticism, which usually provokes a new round of experiments. As a result, the new idea might end up on the scrap heap, perhaps becoming a historical curiosity. Cold fusion seems to have followed this path, and in my own field the suggestion that the two chains of DNA lay side by side instead of intertwined in a double helix.

But once it has passed scrutiny, a new contribution would be absorbed into the edifice of science, expanding and enhancing it, while providing a fragment of immortality to the authors.

My perception was wrong. New scientific ideas can be smothered with silence.

I was aware earlier of the case of Gregor Mendel. His fundamental genetic experiments with peas were ignored for a third of a century. But he had published them in an obscure journal, in an age when meetings and libraries were fewer and journals were circulated by land mail. When his ideas were rediscovered at the start of the twentieth century, Thomas Hunt Morgan set out to disprove them and ended up performing experiments that greatly strengthened their case. A Nobel Prize was Morgan's reward. He wrote in a textbook: "The investigator must . . . cultivate also a skeptical state of mind toward all hypotheses—especially his own—and be ready to abandon them the moment the evidence points the other way."

Morgan's attitude still has a place in science but I no longer believe that it is standard practice. Another strategy has emerged by which some scientists deal with ideas that they dislike. They act as if the discussion or data had never been published, and go about their business without mentioning it.

One example involves the use of a technique called prebiotic synthesis to support the most prevalent idea about the origin of life. This theory proposes that life began on this planet with the accidental formation of an elaborate self-copying molecule, RNA or a close relative. The chemist Graham Cairns-Smith argued in a 1982 book that the technique was flawed and that life's origin by such an event was extremely improbable. He proposed an imaginative alternative. His alternative was debated, but the practice of prebiotic synthesis was continued without discussion.

As I felt that his case was sound, I took it up and extended the arguments against prebiotic synthesis. I published a book and a series of papers in refereed journals, including one devoted entirely to the origin of life. I expected rebuttals and hoped that new controlled experiments would be run that would resolve the issue. The rebuttals did not appear, and citations of my work in the field were sparse. When citations were made, they were usually accompanied by a comment that the RNA-first theory had some problems that were not yet resolved. The resolution would take place by further applications of prebiotic synthesis. A blanket of silence has remained in place in the scientific literature concerning the validity of this technique. Ironically, my ideas have been welcomed by creationists, who advocate a supernatural solution to the origin-of-life problem.

The smother-by-silence practice may be fairly common in science. Professor Kendric Smith of Stanford University has noted a similar pattern in the field of DNA repair, where the contribution of recombination to the repair of damage by ultraviolet radiation has been ignored in key papers. For a moral judgment on this practice, I cannot improve upon Smith's closing quote in his letter to *ASBMB Today*: "In religion one can often be forgiven for one's sins but no one should be forgiven for sins against science."

Artist; composer; recording producer for U2, Talking Heads, Paul Simon, and Coldplay; recording artist; author of A Year of Swollen Appendices

From Revolutionary to Evolutionary

Experimental art and experimental politics have traditionally been convivial bedfellows, though usually, in my opinion, with very little benefit to each other. George Bernard Shaw and his circle fervently supported Stalin against a mounting tide of evidence, the Mitfords supported Hitler, and numerous gifted Italian poets and artists were persuaded by Fascism. Similarly, in the late '60s and early '70s the avant-garde art scene in London was overwhelmed with admiration for Chairman Mao.

As a young artist I was part of that scene, and though never a hard-core Maoist I was impressed by some of his ideas: that intellectuals shouldn't be separated from workers, for example, and that art should somehow serve working-class society. I was sick of "Art for Art's sake" and the insularity of the English art world. I liked, too, the idea that professors should spend a month each year farming, or that designers should find out how it feels to work in a steel foundry. It sounded so benign, from a distance. I felt, like many people at the time, that my society was by comparison stagnant, class-bound, stuck in history, and I admired Mao and the Chinese for their courage in reinventing themselves so dramatically.

Of course, the Americans were saying how dreadful it all was, but I thought, "Well, they would, wouldn't they?" In fact, their criticism increased its credibility, for I believed America had gone fundamentally wrong and her enemies must therefore be my friends. I assumed the U.S. sensed the winds of change issuing from China and was digging her heels in, resisting the future with all her might.

And then, bit by bit, I started to find out what had actually happened, what Maoism meant. I resisted for a while, but I had to admit it: I'd been willingly propagandized, just like Shaw and Mitford and d'Annunzio and countless others. I'd allowed my prejudices to dominate my reason. Those professors working in the countryside were being bludgeoned and humiliated. Those designers were put in the steel foundries as "class enemies"—for the workers to vent their frustrations upon. I started to realize what a monstrosity Maoism had been, and that it had failed in every sense.

Thus began for me a long process of reevaluation. I had to accept that I was susceptible to propaganda, and that propaganda comes from all sides—not just the one I happen to dislike. I realized that I was not by any means a neutral observer, that I came with my own set of prejudices which could be easily tweaked.

I realized too that I had to learn to evaluate opinions separately from those who were giving them: The truth might sometimes come out of a mouth I disliked, but that didn't automatically mean it wasn't the truth.

Maoism, or my disappointment with it, also changed my feelings about how politics should be done. I went from revolutionary to evolutionary. I no longer wanted to see radical change dictated from the top—even if that top claimed to be the bottom, the "voice of the people." I lost faith in the idea that there were quick solutions, that everyone would simultaneously see the light and things would suddenly flip over into a wonderful new reality. I started to believe it was always going to be slow, messy, compromised, unglamorous, bureaucratic, endlessly negotiated—or else extremely dangerous, chaotic, and capricious. In fact, I've lost faith in the idea of ideological politics altogether: I want instead to see politics as the articulation and management of a changing society in a changing world, trying to do a half-decent job for as many people as possible, trying to set things up a little better for the future.

Perhaps this is why I've increasingly come to regard the determinedly nonideological, ecumenical E.U. as the signal political experiment of our time.

PAUL EWALD

Professor of biology, Amherst College; author of Plague Time:
The New Germ Theory of Disease

Received Wisdom

At the end of *The Structure of Scientific Revolutions,* Thomas Kuhn suggested that it is reasonable to trust the general consensus of experts instead of a revolutionary idea, even when the revolutionary idea is consistent with a finding that could not be explained by the general consensus. He reasoned that the general consensus was reached by drawing together countless bits of evidence, and even though it could not explain everything, it had run a gauntlet to which the revolutionary idea had not yet been subjected.

Kuhn's idea seemed sufficiently plausible to lead me to generally trust the consensus of experts in disciplines outside my area of expertise. I still think it is wise to trust the experts when their profession has a good understanding of the processes under consideration. This situation applies to experts on car maintenance, for example, because cars were made by people who shared their knowledge about the function of car parts, and top-notch car mechanics learn this information. It also applies generally to the main principles of mechanical and electrical engineering, biology, physics, and chemistry, because these principles are tested directly or indirectly by countless studies.

I am becoming convinced, however, that the opposite view is often true when the expert opinion pertains to the unknown: The longer the accepted wisdom has been accepted, and the more widespread its acceptance, the more hesitant we should be to trust it, especially if the experts have been studying the question intensively during this period of acceptance and contradictory findings or logic have been presented. The reason is simple. If an explanation has been widely and broadly accepted and convincing evidence still cannot be mustered, then it is quite reasonable to expect that the experts are barking up the wrong, albeit cherished, tree. That is, its acceptance has more to do with the limitations of intellectual ingenuity than with evidence.

This argument provides a clear guideline for allocating trust to experts: Distrust expert opinion in accordance with what is not known about the subject. This guideline is, of course, difficult to apply, because one first has to ascertain whether a discipline actually has valid answers for a given area of inquiry. Consider something as simple as a sprained ankle. Evolutionary considerations suggest that the inflammation and pain associated with sprained ankles are adaptive responses to promote healing and suppressing them would be detrimental to long-term functionality of the joint. I have searched the literature to find out whether any evidence indicates that treatment of sprained ankles with ice, compression, anti-inflammatories, and analgesics promotes or hinders healing and long-term functionality of the joint. In particular, I have been looking for comparisons of treated individuals with untreated controls. I have not found any and am coming to the conclusion that this widely advocated expert opinion is a detrimental holdover from ancient Greek medicine, which often confused the return of the body to a more healthy appearance with the return of the body to a state of health.

More generally, I am coming to the disquieting realization that much of scientific opinion and even more of medical opinion falls into the murky area circumscribed by a lack of adequate knowledge about the processes at hand. This means I must invoke broadly the guideline to distrust expert opinion in proportion to the lack of knowledge in the area. Although this has made me more objectionable, it has also been of great value intellectually and practically, as when, for example, I sprain my ankle.

Professor, University of Vienna; scientific director of the Institute of Quantum Optics and Quantum Information, Austrian Academy of Sciences

Quantum Technology

When journalists asked me twenty years ago what the use of my research was, I proudly told them it had no use whatsoever. I saw an analogue to the "usefulness" of astronomy or of a Beethoven symphony. We didn't do these things, I said, for their usefulness, we did them because they're part of what it means to be human. In the same way, I said, we did basic science—in my case, experiments on the foundations of quantum physics.

It is part of being human to be curious, to want to know more about the world. There are always some of us who are just curious and we follow our nose and investigate with no idea in mind about what it might be useful for. Some of us are even more attracted to a question the more useless it appears. I did my work only because I was attracted both by the mathematical beauty of quantum physics and the counterintuitive conceptual questions it raises. So I told them, up until the early 1990s.

Then came a new and surprising development. The scientific community discovered that the fundamental phenomena of quantum physics had become relevant for more and more novel ways of transmitting and processing information. We now have the completely new field of quantum information science, some of whose basic concepts are quantum cryptography, quantum computation, and even quantum teleportation. All this points us toward a new information technology, in which the same strange fundamental phenomena that attracted me to the field are essential. Quantum randomness makes it possible for us, in quantum cryptography, to send messages so that they are secure. Quantum entanglement, called by Einstein "spooky action at a distance," makes quantum teleportation possible. And quantum computation builds on all counterintuitive features of the quantum world together. When journalists ask me today what the use of my research is, I proudly tell them of my conviction that we will someday have a full quantum information technology, though its specific features are still very much to be developed. So, never claim that your research is "useless."

SETH LLOYD

Quantum-mechanical engineer, Massachusetts Institute of Technology; author of Programming the Universe

What My Students Taught Me

I used to take a dim view of technology. One should live one's life in a simple, low-tech fashion, I thought. No cell phone, keep off the computer, don't drive. No nukes, no remote control, no DVD, no TV. Walk, read, think—that was the proper path to follow.

What a fool I was! A dozen years ago or so, by some bizarre accident, I became a professor of mechanical engineering at MIT. I had never had any training, experience, or education in engineering. My sole claim to engineering expertise was some work on complex systems and a few designs for quantum computers. Quantum-mechanical engineering was in its early days then, though, and MIT needed a quantum mechanic. I was ready to answer the call.

It was not my fellow professors who converted me to technology, über-techno-nerds though they were. Indeed, my colleagues in Mech. E. were by and large somewhat suspicious of me—justifiably so. I was wary of them in turn, as one often is of coworkers who are hugely more knowledgeable than one is oneself. (Outside of the Mechanical Engineering Department, by contrast, I found large numbers of kindred souls: MIT was full of people whose quanta needed fixing, and as a certified quantum mechanic I was glad to oblige.)

No, it was not the brilliant technologists who filled the faculty lunchroom who changed my mind. Rather, it was the students who had come to have me teach them about engineering who taught me to value technology.

Your average MIT undergraduate is pretty technologically adept. In the old days, freshmen used to arrive at MIT having disassembled and reassembled tractors and cars; slightly later on, they arrived having built ham radios and guitar amplifiers; more recently, freshmen and freshwomen were showing up who had a scary facility with computers. Nowadays few of them have used a screwdriver (except maybe to install some more memory in their laptop), but they are eager to learn how robots work and raring to build one themselves.

When I stepped into my first undergraduate classroom, a controls laboratory, I knew just about as little about how to build a robot as the nineteen- and twenty-year-olds who were expectantly waiting for me to teach them how. I was terrified. Within half an hour, the basis for my terror was confirmed. Not only did I know as little as the students, in many cases I knew significantly less; about a quarter of the students knew demonstrably more about robotics than I and were happy to display their knowledge. I emerged from the first lab session a sweaty mess, having managed to demonstrate my ignorance and incompetence in a startling variety of ways.

I emerged from the second lab session a little cooler. There is no better way to learn, and learn fast, than to teach. Humility turns out to have its virtues. It can be rather fun to admit one's ignorance, if that admission takes the form of an appeal to the knowledge of all assembled. As it happened, either through my training in math and physics or through a previous incarnation I possessed more intuitive knowledge of control theory than I had any right to, given my lack of formal education on the subject. Finally, no student is more empowered than the one who has just correctly told her professor that he is wrong and showed him why her solution is the right one.

In the end, the experience of teaching technology I did not know was one of the most intellectually powerful of my life. In my mental ferment of trying to learn the material faster and deeper than my students, I began to grasp concepts and ways of looking at a world of whose existence I had no previous notion. One of the primary features of the lab was a set of analog computers—boxy things festooned with dials and plugs and full of amplifiers, capacitors, and resistors—that were used to simulate, or construct an analog, of the motors and loads we were trying to control. In my feverish attempt to understand analog computers, I constructed a model for a quantum-mechanical analog computer that would operate at the level of individual atoms. This model resulted in one of my best scientific papers.

And technology? Hey, it's not so bad. When it comes to walking in the rain, Gore-Tex and fleece beat oilskin and wool hollow. If we're not going to swamp our world in greenhouse gases, we damn well better design dramatically more efficient cars and power plants. And if I can contribute to technology by designing and helping to build quantum computers and quantum communication systems, so much the better. Properly conceived and constructed technology does not hinder the simple life but helps it.

OK. So I was wrong about technology. What's my next misconception? Religion? God forbid.

ADAM BLY

Founder and editor in chief of Seed

Science's Best Friend

When I started *Seed,* I had a fairly strong aversion to technology. Somehow, sometime, science and technology had become science-and-technology. Two worlds, dissimilar in many respects, likely linked in speech for the practical goal of raising funding and attention for basic research by showing the direct, immediate correlation with usable things. I felt then that the "and" made science more perfunctory and less romantic. And that this was a bad thing on multiple counts.

In the last year, I've come to see the relationship between science and technology very differently. We have reached the point in physics and cosmology, neuroscience, and genetics at least where technology is quintessential to advancement. Technology is not merely making the practice of science faster and less mundane or, as with microscopes, helping us see the otherwise unseeable; it is a distinct yet complementary landscape from which we can advance our knowledge of the natural world.

A physicist at CERN said to me recently that they likely wouldn't have built a new eight-billion-dollar collider if there were a better way of moving the field forward. The Blue Brain

Project is using supercomputers to construct a mind because the neuroscientists involved believe it is the best way of attaining an overall understanding of the brain. Robots, I now appreciate, are not simply novelty items or tools of automation but can be a way of gaining unique insight into humans. From simulation to supercomputing, technology is now (or at least I now see) one of science's very best friends (I could say the same for the arts). And the design, magnitude, and complexity of these technological feats satisfy my (and our) need for romance in our pursuit of truth.

The sum total of all information produced in 2008 will likely exceed the amount of information generated by humans over the past 40,000 years. Science is getting bigger, but as these and other major experiments churn out pentabytes of data, how do we ensure that we are actually learning more? Visualization, and more generally a strong relationship between science and design, will be essential to deriving knowledge from all this information.

PZ MYERS

Biologist, University of Minnesota; blogger, Pharyngula

The Polestar

I always change my mind about everything, and I never change my mind about anything.

That flexibility is intrinsic to being human—more, to being conscious. We are (or should be) constantly learning new things, absorbing new information, and reacting to new ideas, so of course we are changing our minds. In the most trivial sense, learning and memory involve a constant remodeling of the fine details of the brain, and the only time the circuitry will stop changing is when we're dead. In a more profound sense, our major ideas change over time: My five-year-old self, my fifteen-year-old self, and my twenty-five-year-old self were very different people, with different priorities and different understandings of the world, than my current fifty-year-old self. This is simply in the nature of our existence.

In the context of pursuing science, however, there is a substantive context in which we do not change our minds: We have a commitment to following the evidence wherever it leads. We have a kind of overriding metaphysic that says we should set out to find data that will change our minds about a subject—every

295

good research program has as its goal the execution of observations and experiments that will challenge our assumptions—and about that all-important foundation of the scientific enterprise I have never changed my mind, nor can I, without abandoning science altogether.

In my own personal intellectual history, I began my academic career with a focus on neuroscience; I shifted to developmental neurobiology; I later got caught up in developmental biology as a whole; I am now most interested in the confluence of evolution and development. Have I ever changed my mind? I don't think I have in any significant way—I have instead applied a consistent attitude toward a series of problems.

If I embark on a voyage of exploration, and I set as my goals the willingness to follow any lead, pursue any interesting observation, overcome any difficulties, and I end up in some exotic locale that might be very different from my predictions before setting out, have I changed my destination in any way? I would say not; the sine qua non of science is not the conclusions we reach but the process we use to arrive at them, and that is the polestar by which we navigate.

ESTHER DYSON

Chairman, EDventure Holdings; trustee of the Long Now Foundation; author of Release 2.0

Controlling Your Own Data

For a long time, I thought people would start effectively protecting their own privacy online, using tools and services the market would provide. Many companies offered such services, and almost none of them succeeded (at least not with their original business plans). People simply weren't interested: They were both paranoid and careless, and took little trouble to inform themselves. (Of course, if you've ever attempted to read an online privacy statement, you'll understand why.)

But now I've changed my mind and realized that the whole question needs reframing—which Facebook et al are in the process of doing. Users have never learned how to say no to marketers who want their data, but they are getting into the habit of controlling it themselves because Facebook is teaching them that this is a natural thing to do.

Yes, Facebook certainly managed to draw attention to the whole privacy question with its Beacon tracking tool, but for most Facebook users the big question is how many people they can get to see their feed. They are happy to share their information with friends, and they consider it the most natural thing in the world to distinguish among friends (see the Facebook add-on applications such as Top Friends and Cliquey) and to manage their privacy settings to determine who can see which parts of their profile. So why shouldn't they do the same thing vis-à-vis marketers?

For example, I fly a lot, and I use various applications to let certain friends know where I am and where I plan to be. I'd be delighted to share that information with certain airlines and hotels if I knew they would send me special offers. In fact, United Airlines once asked me to send in my frequent-flier statements from up to three competing airlines in exchange for two thousand bonus miles each. I gladly did so, and would have done it for free. I *want* United to know what a good customer I am—and how much more of my business they could win if they offered me even better deals.

In short, for many users the Web is becoming a mirror with users in control, rather than a heavily surveilled stage. The question isn't how to protect users' privacy, but rather, how to give them better tools to control their own data—not by selling privacy or by getting them to "sell" their data but by feeding their natural fascination with themselves and allowing them to manage their own presence. What once seemed like an onerous, weird task becomes akin to self-grooming online.

This begs a lot of questions, I know, including real, coercive invasions of privacy by government agencies, but I think the in-control users of the future will be better equipped to fight the intrusions of overcurious institutions. Give them a little time and a few bad experiences, and they'll start to make the distinction between an airline selling seats and a government that simply won't let you take it off your buddy list.

JARON LANIER

Computer scientist and musician; columnist for Discover *magazine*

VR Therapy?

Here's a happy example of me being wrong.

Other researchers interested in virtual reality were proposing as early as twenty years ago that VR would someday be useful for the treatment of psychological disorders, such as post-traumatic stress disorder.

I did not agree. In fact, I had strong arguments as to why this ought not to work. There was evidence that the brain created distinct "homuncular personas" for virtual-world experiences, and reasons to believe that these personas were tied to increasingly distinct bundles of emotional patterns. Therefore, emotional patterns attached to real-world situations would, I surmised, remain attached to those situations. The earliest research on PTSD treatment in VR seemed awfully shaky to me, and I was not very encouraging to younger researchers who were interested in it.

The idea of using VR for PTSD treatment seemed less likely to work than various other therapeutic applications of VR, which were more centered around somatic processes. For instance, VR can be used as an enhanced physical training environment. The first example, from the 1980s, involved juggling. If virtual juggling balls fly more slowly than real balls, then they are easier to juggle. You can gradually increase the speed in order to provide a more gradual path for improving skills than would be available in physical reality. (This idea came about initially because it was so hard to make early VR systems go as fast as the reality they were emulating. In the old VPL Research lab, where a lot of VR tools were initially prototyped, we were motivated to be alert for potential virtues hiding within the limitations of the era.) Variations on this strategy have become well established. For instance, patients are learning to use prosthetic limbs more quickly by using VR.

Beyond rational argument, I was biased in other ways: The therapeutic use of VR seemed "too cute" and sounded too much like a press-release-in-waiting.

Well, I was wrong. PTSD treatment in VR is now a well-established field, with its own conferences, journals publishing well-repeated results, and clinical practitioners. Sadly, the Iraq War has provided all too many patients and has also prompted increased funding for research in this subfield of VR applications.

One of the reasons I was wrong is that I didn't see that the same tactic we used on juggling balls—of gradually adapting the content and design of a virtual world to the instantaneous state of the user/inhabitant—could be applied in a less somatic way. For instance, in some clinical protocols, a traumatic event is represented in VR with gradually changing levels of realism, as part of the course of treatment.

Maybe I was locked into seeing VR through the filters of the limitations of its earliest years. Maybe I was too concerned about the cuteness factor. At any rate, I'm glad there was a diversity of mindsets in the research community, so that others could see where I didn't.

I'm concerned that diversity of thought in some of the micro-climates of the scientific community is narrowing these days instead of broadening. I blame the nature of certain online tools. Tools like the Wikipedia encourage the false worldview that we already know enough to agree on a single account of reality, and anonymous blog comment rolls can bring out moblike behaviors in young scientists who use them.

At any rate, one of the consolations of science is that being wrong on occasion lets you know you don't know everything and helps renew curiosity. Being aware of being wrong once in a while keeps you young.

AUSTIN DACEY

Philosopher, Center for Inquiry; author of The Secular Conscience

The Power of Reasons

As a teenager growing up in the rural American Midwest, I played in a Christian rock band. We wrote worship songs with texts on religious themes and experienced the jubilation and transport of the music as a visitation by the Holy Spirit. Then one day, as my faith was beginning to waver, I wrote a song with an explicitly nonreligious theme. To my surprise, I discovered that when I performed it I was overcome by the same feelings, and it dawned on me that maybe what we had experienced all along was our own spirit, that what had called to us was the power of music itself.

In truth, I wasn't thinking through much at the time. Later, as a graduate student in philosophy, I did start to think a lot about science and ethics, and I began to undergo a parallel shift of outlook. Having embraced a thoroughly naturalistic, materialistic worldview, I wondered: If everything is just matter, how could anything really matter? How could values be among the objective furniture of the universe? It was not as though anyone expected physicists to discover—alongside electrons, protons, and neutrons—a new fundamental moral particle (the moron?) that would show up whenever people did nice things.

Then there was the late Australian philosopher John Leslie Mackie's famous argument from "queerness": Objective values would have to be such that merely coming to appreciate them would motivate you to pursue them. But given everything we know about how ordinary natural facts behave (they seem to ask nothing of us), how could there possibly be states of affairs with this strange to-be-pursued-ness built into them, and how could we come to appreciate them?

At the same time, I was taken in by the promises found in some early sociobiology that a new evolutionary science of human nature would supplant empty talk about objective values. As philosopher of science Michael Ruse once put it, "Morality is just an aid to survival and reproduction," so "any deeper meaning is illusory." Niceness may seem self-evidently right to us, but things could easily have been the other way around had nastiness paid off more often for our ancestors.

I have since been convinced that I was looking at all of this in the wrong way: Not only are values a part of nature, we couldn't avoid them if we tried.

There is no doubt that had we evolved differently, we would value different things; however, that alone does not show that values are subjective. After all, hearing is accomplished by psychological mechanisms that evolved under natural selection, but it does not follow that the things we hear are any less real. Rather, the reality of the things around us helps to explain why we have the faculty to detect them. The evolved can put us in touch with the objective.

In fact, we are all intimately familiar with entities which are such that to recognize them is to be moved by them. We call them reasons, where a reason is just a consideration that weighs in favor of an action or belief. As separate lines of research by psychologist Daniel Wegner and psychiatrist George Ainslie (as synthesized and interpreted by Daniel Dennett) strongly suggest, our reasons aren't all "in the head," and we cannot help but heed their call.

At some point in our evolution, the behavioral repertoire of our ancestors became complex enough to involve the review and evaluation of numerous possible courses of action and the formation of intentions on the basis of their projected outcomes. In a word, we got options. However, as an ultrasocial species, for whom survival and reproduction depended on close coordination of behaviors over time, we needed to manage these options in a way that could be communicated to our neighbors. That supervisor and communicator of our mental economy is the self—the more-or-less stable "I" that persists through time and feels that it is the author of action. After all, if you want to be able to make reliable threats or credible promises, you need to keep track of who you are, were, and will be. According to this perspective, reasons are a human organism's way of taking responsibility for some of the happenings in its body and environment. As such, they are inherently public and shareable. Reasons are biological adaptations, every bit as real as our hands, eyes, and ears.

I do not expect (and we do not need) a science of good and evil. However, scientific evidence can show how it is that things matter objectively. I cannot doubt without presupposing the power of reasons (for doubting). That cannot be said for the power of the Holy Spirit.

SIMON BARON-COHEN

Psychologist, Autism Research Centre, Cambridge University;
author of Autism and Asperger Syndrome

We Are Not All the Same

When I was young, I believed in equality as a guiding principle in life. It's not such a bad idea, when you think about it. If we treat everyone else as our equals, no one feels inferior—and as an added bonus, no one feels superior. Whereas it's a wonderfully cozy, warm, feel-good idea, I have changed my mind about equality. There seemed to be two moments in my thinking about this principle that revealed some cracks in the perfect idea.

The first moment was in thinking about economic equality. Living on a kibbutz was an interesting opportunity to see that, if you aim for everyone to have exactly the same amount of money or exactly the same possessions or exactly the same luxuries, the only way to achieve this is by legislation. In a small community like a kibbutz, or in an Amish community, where all members of the community can decide on their lifestyles collectively and where the legislation is the result of consensual discussion, economic equality might just be possible.

But in the large towns and cities in which most of us live— and with the opportunities to see how other people live through travel, television, and the Web—it is untenable to expect complete strangers to accept economic equality if it is forced on them. So, for small groups of people who know each other and choose to live together, economic equality might be an achievable principle, but for large groups of strangers, I think we have to admit that it's an unrealistic principle. Economic equality presumes pre-existing relationships based on trust, mutual respect, and choice, which are hard to achieve when you hardly know your neighbors and feel alienated from how your community is run.

The second moment was in thinking about how to square equality with individual differences. Equality is easy to believe in if you believe everyone is basically the same. The problem is that it is patently obvious that we are not all the same. Once you accept the existence of individual differences, this opens the door to some varieties of difference being better than others.

Let's take the thorny subject of gender differences. If males have more testosterone than females, and if testosterone causes not only your beard but also your muscles to grow stronger, it is just naive to cling to the idea that women and men are on a level playing field in competitive sports where strength matters. This is just one example of how individual differences in hormonal levels can play havoc with the idea of biological equality.

Our new research suggests that prenatal testosterone also affects how the mind develops. Higher levels of prenatal testosterone are associated with slower social and language development and reduced empathy. Higher levels of prenatal testosterone are also associated with more autistic traits, stronger interests in systems, and greater attention to detail. A few more drops of this molecule seem to be associated with important differences in how our minds work.

So biology has little time for equality. This conclusion should come as no surprise, since Darwin's theory of evolution was premised on the existence of individual differences upon which natural selection operates. In modern Darwinism, such individual differences are the result of genetic differences — either mutations or polymorphisms in the DNA sequence. Given how hormones and genes (which are not mutually exclusive, genetic differences being one way in which differences in hormone levels come about) can put us onto very different paths in development, how can we believe in equality in all respects?

The other way in which biology is unequal is in the likelihood of developing different medical conditions. Males are sometimes referred to as the weaker sex, because they are more likely to develop a whole host of conditions, among which are autism (four boys for every one girl) or Asperger syndrome (nine boys for every one girl). Given these risks, it becomes almost comical to believe in equality.

I still believe in some aspects of the idea of equality, but I can no longer accept the whole package. The question is, Is it worth holding on to some elements of the idea if you've given up other elements? Does it make sense to have a partial belief in equality? Do you have to either believe in all of it or none of it? My mind has been changed from my youthful starting point, when I hoped that equality could be followed in all areas of life, but I still see value in holding on to some aspects of the principle. Striving to give people equality of social opportunity is still a value system worth defending, even if, in the realm of biology, we have to conclude that equality has no place.

DAVID SLOAN WILSON

Biologist, Binghamton University; author of Evolution for Everyone

Heeding the Butterfly Effect

In 1975, as a newly minted PhD who had just published his first paper on group selection, I was invited by *Science* magazine to review a book by Michael Gilpin titled *Group Selection in Predator-Prey Communities*. Gilpin was one of the first biologists to appreciate the importance of what Stuart Kauffman would call "the sciences of complexity." In his book, he was claiming that complex interactions could make group selection a more important evolutionary force than the vast majority of biologists had concluded on the basis of more simple mathematical models.

Some background: Group selection refers to the evolution of traits that increase the fitness of whole groups, compared with other groups. These traits are often selectively disadvantageous within groups, creating conflict between levels of selection. Group selection requires the standard ingredients of natural selection—a population of groups that vary in their phenotypic properties in a heritable fashion, with consequences for collective survival and reproduction. Standard population-genetics models give the impression that groups are unlikely to vary unless they

308

are initiated by small numbers of individuals with minimal migration among groups during their existence. This kind of reasoning turned group selection into a pariah concept in the 1960s, taught primarily as an example of how not to think. I had become convinced that group selection could be revived for smaller, more ephemeral groups I called "trait groups." Gilpin was suggesting that group selection could also be revived for larger, geographically isolated groups on the basis of complex interactions.

Gilpin focused on the most famous conjecture about group selection, advanced by V. C. Wynne-Edwards in 1962, that animals evolve to avoid overexploiting their resources. Wynne-Edwards had become an icon for everything that was wrong and naive about group selection. Gilpin boldly proposed that animals could indeed evolve to "manage" their resources, based on nonlinearities inherent in predator-prey interactions. As resource exploitation evolves within group selection, there is not a gradual increase in the probability of extinction. Instead, there is a tipping point that suddenly destabilizes the predator-prey interaction, like falling off a cliff. This discontinuity increases the importance of group selection, keeping the predator-prey interaction in the zone of stability.

I didn't get it. To me, Gilpin's model required a house-of-cards of assumptions, a common criticism leveled against earlier models of group selection. I therefore wrote a tepid review of his book. I was probably also influenced by a touch of professional jealousy, as someone who was himself trying to acquire a reputation for reviving group selection.

I didn't get the complexity revolution until I read James Gleick's *Chaos: Making a New Science*, which I regard as one of the best books ever written about science for a general audience. Suddenly I realized that as complex systems, such higher-level biological units as groups, communities, ecosystems, and human cultures would almost certainly vary in their phenotypic properties and that some of this phenotypic variation might be heritable.

Complexity theory became a central theme in my own research.

As one experimental demonstration, William Swenson, who was then my graduate student, created a population of microbial ecosystems by adding one milliliter of pond water from a single, well-mixed source to test tubes containing twenty-nine milliliters of sterilized growth medium. This amount of pond water includes millions of microbes, so the initial variation among the test tubes, based on sampling error, was vanishingly small. Nevertheless, within four days (which amounts to many microbial generations) the test tubes varied greatly in their composition and phenotypic properties, such as the degradation of a toxic compound that had been added to each test tube. Moreover, when the test tubes were selected on the basis of their properties to create a new generation of microbial ecosystems, there was a response to selection. We could select whole ecosystems for their phenotypic properties (in our case, to degrade a toxic compound), in exactly the same way that animal and plant breeders are accustomed to selecting individual organisms.

These results are mystifying, in terms of models that assume simple interactions, but they make perfect sense in terms of complex interactions. Most people have heard about the butterfly effect, whereby an infinitesimal change in initial conditions becomes amplified over the course of time for a complex physical system, such as the weather. Something similar to the butterfly effect was occurring in our experiment, amplifying infinitesimal initial differences among our test tubes into substantial variation over time. A response to selection in the experiments is proof that variation caused by complex interactions can be heritable.

Thanks in large part to complexity theory, evolutionary biologists are once again studying evolution as a multilevel process that can evolve adaptations above the level of individual organisms. I welcome this opportunity to credit Michael Gilpin for the original insight.

NEIL GERSHENFELD

Physicist, Massachusetts Institute of Technology; author of
FAB: The Coming Revolution on Your Desktop—from Personal Computers to Personal Fabrication

Programming Nature

I've long considered myself as working at the boundary between physical science and computer science; I now believe that that boundary is a historical accident and does not really exist.

There's a sense in which technological progress has turned it into a tautological statement. It's now possible to store data in atomic nuclei and use electron bonds as logic gates. In such a computer, the number of information-bearing degrees of freedom is on the same order as the number of physical ones; it's no longer feasible to account for them independently. This means that computer programs can—and, I'd argue, must—look more like physical models, including spatial and temporal variables in the density and velocity of information propagation and interaction. That shouldn't be surprising; the canon of computer science emerged a few decades ago to describe the available computing technology, while the canon of physics emerged a few centuries ago to describe the accessible aspects of nature. Computing technology has changed more than nature has; progress in the former is reaching the limits of the latter.

Conversely, it makes less and less sense to define physical theories by the information technology of the last millennium (a pencil and a sheet of paper); a computational model is every bit as fundamental as one written with calculus. This is seen in frontiers of research in nonlinear dynamics and quantum field theories and black-hole thermodynamics, which look more and more like massively parallel programming models. However, the organization of research has not yet caught up with this content; many of the pioneers doing the work are in neither physics nor computer science departments but are scattered around (and off) campus. Rather than trying to distinguish between programming nature and the nature of programming, I think it makes more sense to recognize not just a technology or theory of information but a single science.

PAUL SAFFO

Technology forecaster

Prediction Engines

When I began my career as a forecaster more than two decades ago, it was a given that the core of futures research lay beyond the reach of traditional quantitative forecasting and its mathematical tools. This meant that futures researchers would not enjoy the full labor-saving benefits of number-crunching computers, but at least it guaranteed job security. Economists and financial analysts might one day wake up to discover that their computer tools were stealing their jobs, but futurists would not see machines muscling their way into the world of qualitative forecasting anytime soon.

I was mistaken. I now believe that in the not too distant future the best forecasters will be not people but machines — ever more capable "prediction engines" probing ever deeper into stochastic spaces. Indicators of this trend are everywhere, from the rise of quantitative analysis in the financial sector to the emergence of computer-based horizon scanning systems in use by governments around the world — and of course the relentless advance of computer systems along the upward-sweeping curve of Moore's Law.

We already have human-computer hybrids at work in the discovery/forecasting space, from Amazon's Mechanical Turk to the myriad online prediction markets. In time, we will recognize that these systems are an intermediate step toward prediction engines, in much the same way that human "computers," who once performed the mathematical calculations on complex projects, were replaced by general-purpose electronic digital computers.

The eventual appearance of prediction engines will also be enabled by the steady uploading of reality into cyberspace, from the growth of Web-based social activities to the steady accretion of sensor data sucked up by an exponentially growing number of devices observing and, increasingly, manipulating the physical world. The result is an unimaginably vast corpus of raw material, grist for the prediction engines as they sift and sort and peer ahead. These prediction engines won't ever exhibit perfect foresight, but as they and the underlying data they work on coevolve, it is a sure bet that they will do far better then mere humans.

ALISON GOPNIK

Psychologist, University of California, Berkeley; coauthor (with Andrew N. Meltzoff and Patricia K. Kuhl) of The Scientist in the Crib: What Early Learning Tells Us About the Mind

Making the Imaginary Real

Recently I've had to change my mind about the very nature of knowledge, because of an obvious but extremely weird fact about children: They pretend all the time. Walk into any preschool and you'll be surrounded by small princesses and superheroes in overalls—three-year-olds spend more waking hours in imaginary worlds than in the real one. Why? Learning about the real world has obvious evolutionary advantages and kids do it better than anyone else. But why spend so much time thinking about wildly, flagrantly, unreal worlds? The mystery about pretend play is connected to a mystery about adult humans, especially vivid for an English professor's daughter like me. Why do we love (obviously false) plays and novels and movies?

The greatest success of cognitive science has been our account of the visual system. There's a world out there sending information to our eyes, and our brains are beautifully designed to recover the nature of that world from that information. I've always thought that science and children's learning worked the same way. Fundamental capacities for causal inference and learning let scientists and children alike get an accurate picture of the world around them—a theory. Cognition was the way we got the world into our minds.

But fiction doesn't fit that picture. It's easy to see why we want the truth, but why do we work so hard at telling lies? I thought that kids' pretend play and grown-up fiction must be a sort of spandrel—a side effect of some other more functional ability. I said as much in a review in *Science* and got floods of e-mail back from distinguished novel-reading scientists. They were all sure fiction was a Good Thing (so was I, of course) but didn't seem any closer than I was to figuring out why.

So the anomaly of pretend play has been bugging me all this time. But finally, trying to figure it out has made me change my mind about the nature of cognition itself.

I still think we're designed to find out about the world, but that's not our most important gift. For human beings, the really important evolutionary advantage is our ability to create new worlds. Look around the room you're sitting in. Every object in that room—the right-angled table, the book, the paper, the computer screen, the ceramic cup—was once imaginary. Not a thing in the room existed in the Pleistocene. Every one of them started out as an imaginary fantasy in someone's mind. And that's even more true of people. All the things I am—a scientist, a philosopher, an atheist, a feminist— all those kinds of people started out as imaginary ideas, too. I'm not making some relativist, postmodern point here; right now, the computer and the cup and the scientist and the feminist are as real as anything can be. But that's just what our human minds do best— take the imaginary and make it real. I think now that cognition is also a way we impose our minds on the world.

In fact, I think now that the two abilities—finding the truth about the world and creating new worlds—are two sides of the same coin. Theories, in science or childhood, don't just tell us what's true, they tell us what's possible and how to get to those possibilities from where we are. When children learn and when they pretend they use their knowledge of the world to create new possibilities. So do we, whether we're doing science or writing novels. I no longer think that Science and Fiction are just both Good Things that complement each other. I think they are the same thing.

Computer scientist, Brandeis University

A Nightmare We Need
to Wake Up From

I've changed my mind about electronic mail. When I first used e-mail, in graduate school in 1980, it was a dream. It was the most marvelous and practical invention of computer science. A text message quickly typed and reliably delivered allowed a new kind of asynchronous communication. It was cheaper (free), faster, reliable, and much more efficient than mail, phone, or fax, making a round-trip in minutes. Colleagues started sharing text-formatted data tables, where fifty kilobytes was a big message.

Then came attachments. This hack to insert 33 percent bloated eight-bit binary files inside of seven-bit text e-mail opened a Pandora's box. Suddenly anyone had the right to send any size package for free, like a Socialist United Parcel Service. Microsoft Outlook made it drag-'n'-drop easy for bureaucrats to send Word documents. Many computer scientists saw the future and screamed "JUST SEND TEXT," but it was too late. Microsoft kept tweaking its proprietary file formats, forcing anyone with e-mail to upgrade Microsoft Office. (I thought they finally stopped with Office 97, but now I am getting DOCX files, which might as well be in Martian.)

We faced AOL newbies, mailing lists, free webmail, Hotmail spam, RTF mail, chain letters, HTML mail, musical mail, Flash mail, JavaScript mail, viruses, spybits, faked URLs, phishing, Nigerian cons, PowerPoint arms races, spam-blocking spam, viral videos, Plaxo updates, Facebook friendings, ad nauseam.

The worst part is the legal precedent that your employers "own" the mail sent out over the network provided. It is as if they own the sound waves that emit from your throat over the phone. An idiot judgment leads to two Kafkaesque absurdities:

First, if you send e-mail with an ethnic slur or receive e-mail with a picture of a naked child or a copyrighted MP3, you can be fired. Use e-mail to organize a union? Fuggeddaboutit!

Second, all e-mail sent and received must now be archived as critical business documents to comply with Sarbanes-Oxley. And Homeland Security wants rights to monitor ISP data streams and stores, and thinks no warrants are needed for data older than ninety days.

Free speech in the Information Age isn't your right to post anonymously on a soapbox blog or newspaper story. It means that if we agree, I should be able to send any data in any file format, with any encryption, from a computer I am using to one you're on, provided we pay for the broadband freight. There is no reason that any government, carrier, or corporation should have any right to store, read, or interpret our digital communications. Show just cause and get a warrant, even if you think an employee is spying or a student is pirating music.

E-mail is now a nightmare we need to wake up from. I don't have a solution yet, but I believe the key to reimagining e-mail is to go fully P2P, to take the corporation and the ISP out of the channel between two people. Our computers are always on, and big enough to run their own SQL database, so we can take back ownership of our personal communications store to the days of private correspondence files, or just letters in a drawer. We can begin with synchronous messaging (both sender and receiver are online and agree to a transfer of information)—a cross between file sharing, SMS texting, and instant messaging—and then add identity management, grid backup mechanisms, asynchronous delivery, multiple recipients, and of course reliability.

Until then, just call me.

Science writer; author of The Ten Most Beautiful Experiments

Electricity in the Raw

I used to think that the most fascinating thing about physics was theory and that the best was still to come. But as physics has grown vanishingly abstract, I've been drawn in the opposite direction, to the great experiments of the past.

First I determined to show myself that electrons really exist. Firing up a beautiful old apparatus I found on eBay—a bulbous vacuum tube big as a melon mounted between two coils—I replayed J. J. Thomson's famous experiment of 1897, in which he measured the charge-to-mass ratio of an electron beam. It was thrilling to see the bluish-green cathode ray dive into a circle as I energized the electromagnets. Even better, when I measured the curve and plugged all the numbers into Thomson's equation, my answer was off by only a factor of two. Pretty good for a journalist. I had less success with the stubborn Millikan oil-drop experiment. Mastering it, I concluded, would be like learning to play the violin.

Electricity in the raw is as mysterious as superstrings. I turn down the lights and make my Geissler tubes glow with the touch of a high-voltage wand energized by a brass-and-mahogany Ruhmkorff coil. I coax the ectoplasmic rays in my de la Rive tube to rotate around a magnetized pole.

Maybe in a year or two, the Large Hadron Collider will make this century's physics interesting again. Meanwhile, as soon as I find a nice spinthariscope, I'm ready to go nuclear.

GEOFFREY MILLER

Evolutionary psychologist, University of New Mexico; author of The Mating Mind: How Sexual Choice Shaped the Evolution of Human Nature

Out Among the People

Guys lost on unfamiliar streets often avoid asking for directions from locals. We try to tough it out with map and compass. Admitting that you're lost feels like admitting that you're stupid. This is a stereotype, but it has a large grain of truth. It's also a good metaphor for a big overlooked problem in the human sciences.

We're trying to find our way around the dark continent of human nature. We scientists are being paid to be the bus-tour guides for the rest of humanity. They expect us to know our way around the human mind, but we don't.

So we try to fake it, without asking the locals for directions. We try to find our way from first principles of geography (theory) and from maps of our own making (empirical research). The roadside is crowded with locals, and their brains are crowded with local knowledge, but we are too arrogant and embarrassed to ask the way. Besides, they look strange and might not speak our language. So we drive around in circles, inventing and rejecting successive hypotheses about where to find the scenic vistas that would entertain and enlighten the tourists (laypeople, a.k.a. taxpayers). Eventually, our busload starts grumbling about tour-guide rip-offs in boring countries. We drive faster, make more frantic observations, promise magnificent sights just around the next bend.

I used to think this was the best we could do as behavioral scientists. I figured that the intricacies of human nature were not just dark but depopulated—that a few exploratory novelists and artists had sought the sources of our cognitive Amazons and emotional Niles but that nobody actually lived there.

Now I've changed my mind: There are local experts about almost all aspects of human nature, and the human sciences should find their way by asking them for directions. These locals are the thousands or millions of bright professionals and practitioners in thousands of different occupations. They are the people who went to our high schools and colleges but who found careers with higher pay and shorter hours than academic science offers. Almost all of them know important things about human nature that behavioral scientists have not yet described, much less understood. Marine drill sergeants know a lot about aggression and dominance. Master chess players know a lot about if/then reasoning. Prostitutes know a lot about male sexual psychology. Schoolteachers know a lot about child development. Trial lawyers know a lot about social influence. The dark continent of human nature is already richly populated with autochthonous tribes, but we scientists don't bother to talk to these experts.

My suggestion is that whenever we try to understand human nature in some domain, we should identify several groups of people who are likely to know a lot about that domain already, from personal, practical, or professional experience. We should seek out the most intelligent, articulate, and experienced locals—the veteran workers, managers, and trainers. Then, we should talk with them, face-to-face, expert-to-expert, as collaborating peers, not as researchers "running subjects" or "interviewing informants." We may not be able to reimburse them at their professional hourly wage, but we can offer other forms of prestige, such as coauthorship on research papers.

For example, suppose a psychology PhD student wants to study emotional adaptations, such as fear and panic, that evolved for avoiding predators. She learns about the existing research (mostly by Clark Barrett at UCLA) but doesn't have any great ideas for her dissertation research. The usual response is three years of depressed soul-searching, random speculation, and fruitless literature reviews. This phase of idea-generation could progress much more happily if she just picked up the telephone and called some of the people who spend their whole professional lives thinking about how to induce fear and panic. Anyone involved in horror-movie production would be a good start: screenwriters, monster designers, special-effects technicians, directors, and editors. Other possibilities would include talking with:

- Halloween mask designers
- horror-genre novelists
- designers of "first person shooter" computer games
- clinicians specializing in animal phobias and panic attacks
- Kruger Park safari guides
- circus lion tamers
- dogcatchers
- bullfighters
- survivors of wild-animal attacks
- zookeepers who interact with big cats, snakes, and raptors

A few hours of chatting with such folks would probably be more valuable in sparking some dissertation ideas than months of library research.

The division of labor generates wondrous prosperity and an awesome diversity of knowledge about human nature in different occupations. Psychology could continue trying to rediscover all this knowledge from scratch. Or it could learn some humility and start listening to the real expertise about human nature already acquired by every bright worker in every factory, office, and mall.

Science editor of The Independent

Bleak

I was born in the second half of the twentieth century and for most of my life I grew up in the perhaps naive belief that the twenty-first would be somehow better, shinier, and brighter than the last. We even used it as a positive adjective: "twenty-first-century health care," a "twenty-first-century car," even a "twenty-first-century way of life." Over the past decade or so, my opinion has gradually changed. I now believe that however bad the twentieth century was—and it brought us the horrors of the Holocaust and nuclear proliferation—the new century will be far worse.

Writing about science as a career takes you on an extraordinary journey of progression, giving you the illusion that everything is on an unfaltering course of improvement. Many other specialisms in daily journalism—politics, arts, legal affairs, crime, education—seem to follow a circular path of reporting: The same type of stories appear time and time again. But science is all about standing on the shoulders of the giants who came before you; the inverted pyramid of scientific knowledge continues its exponential growth. Thus it seems evident that things can only get better, as more questions can be answered and more problems solved.

History, too, supports the idea of a progressively better world. Vaccines, drugs, better hygiene and housing, clean water, and other general improvements in health and well-being are now taken for granted. Today people in developed countries live longer and healthier lives than any previous generation—and often without the pain that went with living in the age before science. Anyone who doubts the improvements in medical science should read Claire Tomalin's biography of Samuel Pepys, in which she describes in some detail how surgeons removed a bladder stone through his penis without the benefit of anesthetic. Amazingly, he survived.

But as the first decade of this century enters its final years, my optimism with regard to the remaining nine has waned. I no longer see the phrase "twenty-first century" as synonymous with progress. There is no single event or fact that has led to this change of mind; if pressed, I would blame two interacting phenomena—global warming and the inexorable growth in the human population.

This century will see both effects come into deadly play. By mid-century, there will be half again as many people on the planet as there are now—some nine billion or more—and the resources available to support them will be severely degraded, even without climate change. But we know that the world will be warmer, perhaps significantly so by mid- to late century, and this will put intolerable pressure on the only life-support system we know—Planet Earth.

I have also changed my mind about the assessments of the Intergovernmental Panel on Climate Change. They are much too conservative and have underestimated the future impact of melting polar ice sheets and rising sea levels. The biggest influence on changing my mind on this point has been James Hansen, director of the Goddard Institute for Space Studies, who in 2007 coauthored a twenty-nine-page scientific paper in the *Philosophical Transactions of the Royal Society* detailing why the scale of the threat has put Earth in imminent peril. Hansen believes that nothing short of a planetary rescue will save us from global environmental cataclysm and that we have less than ten years to act.

The sea ice of the Arctic is melting far faster than anyone had predicted, and the record minimum seen in summer 2007 (which followed the previous record minimum of 2005) has shocked even the most seasoned Arctic observers. The stability of the Greenland ice sheet and the West Antarctic ice sheet in the Southern Hemisphere, both of which have the potential to raise sea levels by many meters, is far more precarious than any IPCC report has hitherto suggested. Given that many hundreds of millions of people live within a few meters of sea level, and many of them are already competing for ever more limited supplies of freshwater, the issue of impending sea-level rise will become one of the most pressing problems facing humanity this century.

Added to this is the issue of positive feedbacks within the climate system—the factors that will make climate change far worse as carbon dioxide levels continue to rise. As Hansen and others have pointed out, there seem to be many more positive reinforcers of climate change than the negative feedbacks that might possibly help to limit the damage. In short, we are tinkering with a global climate system that could go dangerously out of control and at a far faster rate than anyone has imagined as they peer into the crystal balls of their computer models. If it happens at all, the positive feedbacks will begin to exert their global influence early in the twenty-first century.

James Lovelock, the veteran Earth scientist and inventor of the Gaia theory, has said that the four horsemen of the apocalypse will ride again this century as climate change triggers a wave of mass migrations, pandemics, and violent conflicts. I would very much like to believe he is wrong, that we can somehow act in international unison as a common federation of humanity to address overpopulation and climate change. I wish I could believe that we have the resolve to tackle the two issues that could end the civilized progress of science and culture. Unfortunately, at this moment in time, I'm not ready to change my mind on that.

ROGER BINGHAM

Cofounder and director of the Science Network; neuroscience researcher, Center for Brain and Cognition, University of California, San Diego; creator of PBS science programs; coauthor (with Peggy La Cerra) of The Origin of Minds: Evolution, Uniqueness, and the New Science of the Self

Changing My Religion

I was once a devout member of the Church of Evolutionary Psychology.

I believed in modules—lots of them. I believed that the mind could be thought of as a confederation of hundreds, possibly thousands, of information-processing neural adaptations. I believed that each of these mental modules had been fashioned by the relentless winnowing of natural selection as a solution to problems encountered by our hunter-gatherer ancestors in the Pleistocene. I believe I actually said that we were living in the Space Age with brains from the Stone Age. Which was clever—but not, it turned out, particularly wise.

Along with the Church Elders, I believed that this was our universal evolutionary heritage: that if you added together a whole host of these domain-specific minicomputers—a face-recognition module, a spatial-relations module, a rigid-object-mechanics module, a tool-use module, a social-exchange module, a child-care module, a kin-oriented motivation module, a sexual-attraction module, a grammar-acquisition module, and so on—then you had the neurocognitive architecture that comprises the human mind. Along with them, I believed that what made the human mind special was not fewer of these "instincts" but more of them.

I was so enchanted by this view of life that I used it as the conceptual scaffolding upon which to build a multimillion-dollar critically acclaimed PBS series that I created and hosted in 1996.

And then I changed my mind.

Actually, I prefer to say that I experienced a conversion. My conversion (literally, a turning around), my adoption of new beliefs, was prompted primarily by conversations—first and foremost with Peggy La Cerra, an apostate from the Church of Evolutionary Psychology's inner sanctum, then with a group of colleagues including neuroscientists, evolutionary biologists, and philosophers. Two years later, La Cerra and I published, in the *Proceedings of the National Academy of Sciences*, an alternative model of the mind, then followed that with a book in 2002.

Although this is not the place to detail the arguments, we suggested that the selective pressures of navigating ancestral environments—particularly the social world—would have required an adaptively flexible, online information-processing system and would have driven the evolution of the neocortex. We claimed that the ultimate function of the mind is to devise behavior that wards off the depredations of entropy and keeps our energy bank balance in the black. So our universal evolutionary heritage is not a bundle of instincts but a self-adapting system responsive to environmental stimuli, constantly analyzing bioenergetic costs and benefits, creating a customized database of experiences and outcomes, and generating minds that are unique by design.

We also explained the construction of selves, how our systems adapt to different "marketplaces," and the importance of reputation effects—a richly nuanced story, which explains why the phrase "I changed my mind" is, with all due respect, the kind of rather simplistic folk-psychological language I hope we will eventually clean up. I think it was Mallarmé who said it was the duty of the poet to purify the language of the tribe. That task now falls also to the scientist.

This model of the mind that I have now subscribed to for about a decade is the bible at the Church of Theoretical Evolutionary Neuroscience (of which I am a cofounder). It was created in alignment with both the adaptationist principles of evolutionary biologists and psychologists (who, at the time, tended to pay little attention to the actual workings of the brain at the implementation level of neurons) and the constructivist principles of neuroscientists (who tended to pay little attention to adaptationism). It would be unrealistic, however, to claim that the two perspectives have yet been satisfactorily reconciled.

And this time, I am not so devout.

Some Evolutionary Psychologists promoted their ideas with a fervor that has been described as evangelical. To a certain extent, that seems to go with the evolutionary territory. Think of the ideological feuds surrounding sociobiology, the renewed debates about levels of selection, and so on. Of course, it could be argued that the latest subfields of neuroscience, such as neuroeconomics and social cognitive neuroscience, are not immune to these enthusiasms (the word comes from the Greek *enthousiasmos*: inspired or possessed by a god or gods). Think of the fMRI-mediated neophrenological explosion of areas said to be the neural correlate of some characteristic or other; or whether the mirror-neuron system can possibly carry all the conceptual freight currently being assigned to it.

Even in science, a seductive story will sometimes, at least for a while, outpace the data. Maybe that's inevitable in the pioneering phase of a fledgling discipline. But that's when caution is most necessary—when the engine of discovery is running more on faith than facts. That's the time to remember that hubris is a sin in science as well as religion.

Evolutionary biologist, Charles Simonyi Professor for the Public Understanding of Science, University of Oxford; author of The God Delusion

The Handicap Principle

When a politician changes his mind, he is a "flip-flopper." Politicians will do almost anything to disown the virtue—as some of us might see it—of flexibility. Margaret Thatcher said, "The lady is not for turning." Tony Blair said, "I don't have a reverse gear." Candidates in the recent Democratic presidential primaries whose original decision to vote in favor of invading Iraq had been based on information believed in good faith but later known to be false stood by their earlier error for fear of the dread accusation "flip-flopper."

How very different is the world of science. Scientists actually gain kudos by changing their minds. If a scientist cannot come up with an example of having changed his mind during his career, he is hidebound, rigid, inflexible, dogmatic! It is not really all that paradoxical, when you think about it, that prestige in politics and science should move in opposite directions.

I have changed my mind, as it happens, about a highly paradoxical theory of prestige in my own field, evolutionary biology.

That theory is the Handicap principle, suggested by the Israeli zoologist Amotz Zahavi. I thought it was nonsense and said so in my first book, *The Selfish Gene*. In the second edition I changed my mind, as the result of some brilliant theoretical modeling by my Oxford colleague Alan Grafen.

Zahavi originally proposed his Handicap principle in the context of sexual advertisement by male animals to females. The long tail of a cock pheasant is a handicap. It endangers the male's survival. Other theories of sexual selection reasoned—plausibly enough—that the long tail is favored in spite of that. Zahavi's maddeningly contrary suggestion was that females prefer long-tailed males not despite the handicap but precisely because of it. To use Zahavi's preferred style of anthropomorphic whimsy, the male pheasant is saying to the female, "Look what a fine pheasant I must be, for I have survived in spite of lugging this incapacitating burden around behind me."

For Zahavi, the handicap has to be a genuine one, authentically costly. A fake burden—the equivalent of the padded shoulder as counterfeit of physical strength—would be rumbled by the females. In Darwinian terms, natural selection would favor females who scorn padded males and choose instead males who demonstrate genuine physical strength in a costly, and therefore unfakeable, way.

Zahavi generalized his theory from sexual selection to all spheres in which animals communicate with one another. He himself studies Arabian Babblers, little brown birds of communal habit, who often "altruistically" feed one another. Conventional "selfish gene" theory would seek an explanation in terms of kin selection or reciprocation. Indeed, such explanations are usually right (I haven't changed my mind about that). But Zahavi noticed that the most generous babblers are the socially dominant individuals, and he interpreted this in handicap terms. Translating, as ever, from bird to human language, he put it into the mouth of a donor bird like this: "Look how superior I am to you. I can even afford to give you food." Similarly, some individuals act as "sentinels," sitting conspicuously in a high tree and not feeding, watching for hawks and warning the rest of the flock, who are therefore able to get on with feeding. Again eschewing kin selection and other manifestations of conventional selfish genery, Zahavi's explanation followed his own paradoxical logic: "Look what a great bird I am. I can afford to risk my life sitting high in a tree watching out for hawks, saving your miserable skins for you and allowing you to feed while I don't." What the sentinel pays out in personal cost he gains in social prestige, which translates into reproductive success. Natural selection favors conspicuous and costly generosity.

You can see why I was skeptical. It is all very well to pay a high cost to gain social prestige. Maybe the raised prestige does indeed translate into Darwinian fitness. But the cost itself still has to be paid, and that will wipe out the fitness gain. Don't evade the issue by saying that the cost is only partial and will only partially wipe out the fitness gain. After all, won't a rival individual come along and outcompete you in the prestige stakes by paying a greater cost? And won't the cost therefore escalate until the point where it exactly wipes out the alleged fitness gain?

Verbal arguments of this kind can take us only so far. Mathematical models are needed, and various people supplied them, notably John Maynard Smith, who concluded that Zahavi's idea, though interesting, just wouldn't work. Or, to be more precise, Maynard Smith couldn't find a mathematical model that led to the conclusion that Zahavi's theory might work. He left open the possibility that somebody else might come along later with a better model. That is exactly what Alan Grafen did, and now we all have to change our minds.

I translated Grafen's mathematical model back into words in the second edition of *The Selfish Gene* (pp. 309–13), and I shall not repeat myself here. In one sentence, Grafen found an evolutionarily stable combination of male advertising strategy and female credulity strategy that turned out to be unmistakably Zahavian. I was wrong to dismiss Zahavi, and so were a lot of other people.

Nevertheless, a word of caution. Grafen's role in this story is of the utmost importance. Zahavi advanced a wildly paradoxical and implausible idea, which—as Grafen was able to show— eventually turned out to be right. But we must not fall into the trap of thinking that therefore the next time somebody comes up with a wildly paradoxical and implausible idea, it too will turn out to be right. Most implausible ideas are implausible for a good reason. Although I was wrong in my skepticism and I have now changed my mind, I was still right to have been skeptical in the first place. We need our skeptics, and we need our Grafens to go to the trouble of proving them wrong.

Novelist; physicist, University of California, Irvine; author of Timescape

A Law of Laws

Richard Feynman held that philosophy of science is as useful to scientists as ornithology is to birds. Often this is so. But the unavoidable question about physics is, Where do the laws come from?

Einstein hoped that God had no choice in making the universe. But philosophical issues seem unavoidable when we hear of the "landscape" of possible string-theory models. As now conjectured, the theory leads to 10,500 solution universes—a horrid violation of Occam's Razor that we might term "Einstein's nightmare."

I once thought that the laws of our universe were unquestionable, in that there was no way for science to address the question. Now I'm not so sure. Can we hope to construct a model of how laws themselves arise?

Many scientists dislike even the idea of doing this, perhaps because it's hard to know where to start. Perhaps ideas from the currently chic technology, computers, are a place to start. Suppose we treat the universe as a substrate carrying out computations, a metacomputer.

Suppose that precise laws require computation, which can never be infinitely exact. Such a limitation might be explained by counting the computational capacity of a sphere around an "experiment" that tries to measure outcomes of those laws. The sphere expands at the speed of light, say, so longer experiment times give greater precision. Thinking mathematically, this sets a limit on how sharp differentials can be in our equations. A partial derivative of time cannot be better than the time to compute it.

In a sense, there may be an ultimate limit on how well known any law can be, especially one that must describe all of spacetime, like classical relativity. It can't be better than the total computational capacity of the universe, or the capacity within the light sphere we can see.

I wonder if this idea can somehow define the nature of laws, beyond the issue of their precision. For example, laws with higher derivatives will be less descriptive because their operations cannot be carried out in a given volume over a finite time.

Perhaps the infinite discreteness required for formulating any mathematical system could be the limiting bound on such discussions. There should be energy bounds, too, within a finite volume, and thus limits on processing power set by the laws of thermodynamics. Still, I don't see how these arguments tell us enough to derive, say, general relativity.

Perhaps we need more ideas to derive a Law of Laws. Can we use the ideas of evolution? Perhaps invoke selection among laws that penalize those laws that lead to singularities—and thus take those regions of spacetime out of the game? Lee Smolin tried a limited form of this by supposing universes reproduce through black-hole collapses. Ingenious, but that didn't seem to lead very far. He imagined some variation in reproduction of budded-off generations of universes, so their fundamental parameters varied a bit. Then selection could work.

In a novel of a decade ago, *Cosm*, I invoked intelligent life, rather than singularities, to determine selection for universes that can foster intelligence, as ours seems to. (I didn't know about Lee's ideas at the time.) The idea is that a universe hosting intelligence evolves creatures that find ways in the laboratory to make more universes, which bud off and can further engender more intelligence, and thus more experiments that make more universes. This avoids the problem of how the first universe started, of course. Maybe the Law of Laws could answer that, too?

LERA BORODITSKY

Assistant professor, Department of Psychology, Stanford University

Operation Perceptual Freedom

I thought that languages and cultures shape the ways we think, I suspected they shaped the ways we reason and interpret information, but I didn't think languages could shape the nuts and bolts of perception—the way we see the world. That part of cognition seemed too low-level, too hard-wired, too constrained by the constants of physics and physiology to be affected by language.

Then studies started coming out claiming to find cross-linguistic differences in color memory. For example, it was shown that if your language makes a distinction between blue and green (as in English), then you're less likely to confuse a blue color chip for a green one in memory. In a study like this, you would see a color chip, it would then be taken away, and after a delay you would have to decide whether another color chip was identical to the one you saw or not.

Of course, showing that language plays a role in memory is different from showing that it plays a role in perception. Things often get confused in memory, and it's not surprising that people

342

may rely on information available in language as a second resort. But that wouldn't mean that speakers of different languages actually saw colors differently as they were looking at them. I thought that if you made a task where people could see all the colors as they were making their decisions, then there wouldn't be any cross-linguistic differences.

I was so sure that language couldn't shape perception that I designed a set of experiments to demonstrate this. In my lab, we jokingly referred to this line of work as "Operation Perceptual Freedom." Our mission: to free perception from the corrupting influences of language.

We did one experiment after another, and each time, to my surprise and annoyance, we found consistent cross-linguistic differences. They were there even when people could see all the colors at the same time when making their decisions. They were there even when people had to make objective perceptual judgments. They were there when no language was involved or necessary in the task at all. They were there when people had to reply very quickly. We kept seeing cross-linguistic differences over and over again, and the only way to get them to go away was to disrupt the language system. If we stopped people from being able to fluently access their language, then the cross-linguistic differences in perception went away.

I had set out to show that language didn't affect perception, but I found exactly the opposite. It turns out that languages meddle in very low-level aspects of perception and without our knowledge or consent shape the very nuts and bolts of how we see the world.

GEORGE B. DYSON

Science historian; author of Project Orion: The True Story of the Atomic Spaceship

The Russian Colonization of North America

Russians arrived on the western shores of North America after crossing their eastern ocean in 1741. After an initial period of exploration, they settled down for a full century, until relinquishing their colonies to the United States. From 1799 to 1867, the colonies were governed by the Russian-American Company, a for-profit monopoly chartered under the deathbed instructions of Catherine the Great.

The Russian-American period has been treated unkindly by historians from both sides. Soviet-era accounts, though acknowledging the skill and courage of Russian adventurers, saw this Tsarist experiment at building a capitalist, American society as fundamentally flawed, casting the native Aleuts as exploited serfs. American accounts, glossing over our own subsequent exploitation of Alaska's indigenous population and natural resources, sought to emphasize that we liberated Alaska from Russian overseers who were worse and would never be coming back.

Careful study of primary sources has convinced me that these interpretations are not supported by the facts. The Aleutian archipelago was a spectacularly rich environment with an unusually dense, thriving population whose physical and cultural well-being was devastated by contact with European invaders. But as permanent colonists, the Russians were not so bad. The results were closer to the settlement of Greenland by Denmark than to our own settlement of the American West.

Although during the initial decades leading up to the consolidation of the Russian-American Company there was sporadic conflict (frequently disastrous to the poorly armed and vastly outnumbered Russians) with the native population, the colonies soon entered a relatively stable state based on cooperation, intermarriage, and official policies that provided social status, education, and professional training to children of mixed Aleut-Russian birth. Within a generation or two, the day-to-day administration of the Russian-American colonies was largely in the hands of native-born Alaskans. As exemplified by the Russian adoption and adaptation of the Aleut kayak, or baidarka, many indigenous traditions and technologies (including sea-otter-hunting techniques and the working of native copper deposits) were adopted by the new arrivals, reversing the usual trend in colonization, when indigenous technologies are replaced.

The Russians instituted public education, preservation of the Aleut language through transliteration of religious and other texts into Aleut via an adaptation of the Cyrillic alphabet, vaccination of the native population against smallpox, and science-based sea-mammal conservation policies that were far ahead of their time. There were no such things as reservations for the native population in Russian America, and we owe as much to the Russians as to the Alaska Native Claims Settlement Act of 1971 that this remains true today.

The lack of support for the colonies by the home government (St. Petersburg was half a world away, and Empress Catherine's instructions a fading memory) eventually forced the sale to the United States, but also necessitated the resourcefulness and local autonomy that made the venture a success.

Russian America was a social and technological experiment that worked, until political compromises brought the experiment to a halt.

Professor of psychology, provost and senior vice president, Tufts University

Stretching Your Mind

I used to believe that the paramount purpose of a liberal education was threefold:

1. Stretch your mind. Reach beyond your preconceptions. Learn to think of things in ways you have never thought before.
2. Acquire tools with which to critically examine and evaluate new ideas, including your own cherished ones.
3. Settle eventually on a framework or set of frameworks that organize what you know and believe and that guide your life as an individual and a leader.

I still believe #1 and #2. I have changed my mind about #3. I now believe in a new version of #3, which replaces the above with the following:

A. Learn new frameworks, and be guided by them.
B. But never get so comfortable as to believe that your

frameworks are the final word, recognizing the strong psychological tendencies that favor sticking to your worldview. Learn to keep stretching your mind, keep stepping outside your comfort zone, keep trying to put yourself in the shoes of others whose frameworks or cultures are alien to you, and have an open mind to different ways of parsing the world. Before you critique a new idea or another culture, master it to the point at which its proponents or members recognize that you get it.

Settling into a framework is easy. The brain is built to perceive the world through structured lenses—cognitive scaffolds on which we hang our knowledge and belief systems.

Stretching your mind is hard. Once we've settled on a worldview that suits us, we tend to hold on. New information is bent to fit, information that doesn't fit is discounted, and new views are resisted.

By "framework" I mean any one of a range of conceptual or belief systems—either explicitly articulated or implicitly followed. These include narratives, paradigms, theories, models, schemas, frames, scripts, stereotypes, and categories; they include philosophies of life, ideologies, moral systems, ethical codes, worldviews, and political, religious, or cultural affiliations. These are all systems that organize human cognition and behavior by parsing, integrating, simplifying, or packaging knowledge or belief. They tend to be built on loose configurations of seemingly core features, patterns, beliefs, commitments, preferences, or attitudes that have a foundational and unifying quality in one's mind or in the collective behavior of a community. When they involve the perception of people (including oneself), they foster a sense of affiliation that may trump essential features or beliefs.

What changed my mind was the overwhelming evidence of biases in favor of perpetuating prior worldviews. The brain maps information onto a small set of organizing structures, which serve as cognitive lenses, skewing how we process or seek new information. These structures drive a range of phenomena, including the perception of coherent patterns (sometimes where none exist), the perception of causality (sometimes where none exists), and the perception of people in stereotyped ways.

Another family of perceptual biases stems from our being social animals (even scientists are social animals!), susceptible to the dynamics of in-group versus out-group affiliation. A well-known bias of group membership is the overattribution effect, according to which we tend to explain the behavior of people from other groups in dispositional terms ("That's just the way they are") but our own behavior in much more complex ways, including a greater consideration of the circumstances. Group attributions are also asymmetrical with respect to good versus bad behavior. For groups that you like, including your own, positive behaviors reflect inherent traits ("We're basically good people") and negative behaviors are either blamed on circumstances ("I was under a lot of pressure") or discounted ("Mistakes were made"). In contrast, for groups that you dislike, negative behaviors reflect inherent traits ("They can't be trusted") and positive behaviors reflect exceptions ("He's different from the rest"). Related to attribution biases is the tendency—perhaps based on having more experience with your own group—to believe that individuals within another group are similar to one another ("They're all alike") whereas your own group contains a spectrum of individuals, including "a few bad apples." When two groups accept bedrock commitments that are fundamentally opposed, the result is conflict—or war.

Fortunately, the brain has other systems that allow us to counteract these tendencies to some extent. This requires conscious effort, the application of critical reasoning tools, and practice. The plasticity of the brain permits change—within limits.

To assess genuine understanding of an idea one is inclined to resist, I propose a version of Turing's Test tailored for this purpose: You understand something you are inclined to resist only if you can fool its proponents into thinking you get it. Few critics can pass this test. I would also propose a cross-cultural Turing Test for would-be cultural critics (a Golden Rule of cross-group understanding): Before critiquing a culture or aspect thereof, you should be able to navigate seamlessly within that culture as judged by members of that group.

By rejecting #3, you give up certainty. Certainty feels good, and it is a powerful force in leadership. The challenge, as Bertrand Russell put it in *The History of Western Philosophy*, is "to teach how to live without certainty, and yet without being paralyzed by hesitation."

DENIS DUTTON

*Professor of the philosophy of art, University of Canterbury,
New Zealand; author of* The Art Instinct: Beauty, Pleasure,
and Human Evolution

Sexual Selection: A Revived Teleology

The appeal of Darwin's theory of evolution—and the horror of it, for some theists—is that it expunges from biology the concept of purpose, of teleology, thereby converting biology into a mechanistic, canonical science. In this respect, the author of *The Origin of Species* may be said to be the combined Copernicus, Galileo, and Kepler of biology. Just as these astronomers gave us a view of the heavens in which no angels were required to propel the planets in their orbs and the Earth was no longer the center of the celestial system, so Darwin showed that no God was needed to design the spider's intricate web and that man is in truth but another animal.

That's how the standard story goes, and it is pretty much what I used to believe, until I read Darwin's later book, *The Descent of Man*, his treatise on the evolution of the mental life of animals, including the human species. This is the work in which Darwin introduces one of the most powerful ideas in the study of human nature, one that can explain why the capacities of the human mind so extravagantly exceed what would have been required for hunter-gatherer survival on the Pleistocene savannahs. The idea is sexual selection, the process by which men and women in the Pleistocene chose mates according to varied physical and mental attributes and in so doing "built" the human mind and body as we know it.

In Darwin's account, human sexual selection comes out looking like a kind of domestication. Just as human beings domesticated dogs and alpacas, roses and cabbages, through selective breeding, they also domesticated themselves as a species through the long process of mate selection. Describing sexual selection as human self-domestication should not seem strange. Every direct prehistoric ancestor of every person alive today at times faced critical survival choices: whether to run or hold ground against a predator, which road to take toward a green valley, whether to slake an intense thirst by drinking from a brackish pool. These choices were frequently instantaneous and intuitive and, needless to say, our direct ancestors were the ones with the better intuitions.

However, there was another kind of crucial intuitive choice faced by our ancestors: whether to choose this man or that woman as a mate with whom to rear children and share a life of mutual support. It is inconceivable that decisions of such emotional intimacy and magnitude were not made with an eye toward the character of the prospective mate, and that these decisions did not therefore figure in the evolution of the human personality—with its tastes, values, and interests. Our direct ancestors, male and female, were the ones who were chosen by each other.

Darwin's theory of sexual selection has disquieted and irritated many otherwise sympathetic evolutionary theorists because, I suspect, it allows purposes and intentions back into evolution through an unlocked side door. The slogan memorized by generations of students of natural selection is *random mutation and selective retention*. The "retention" in natural selection is strictly nonteleological, a matter of brute, physical survival. The retention process of sexual selection, however, is, with human beings, in large measure purposive and intentional. We may puzzle about whether, say, peahens have "purposes" in selecting peacocks with the largest tails. But other animals aside, it is absolutely clear that with the human race, sexual selection describes a revived evolutionary teleology. Though it is directed toward other human beings, it is as purposive as the domestication of those wolf descendants that became familiar household pets.

Every Pleistocene man who chose to bed, protect, and provision a woman because she struck him as, say, witty and healthy, and because her eyes lit up in the presence of children, along with every woman who chose a man because of his hunting skills, fine sense of humor, and generosity, was making a rational, intentional choice that in the end built much of the human personality as we now know it.

Darwinian evolution is therefore structured across a continuum. At one end are purely natural selective processes that give us, for instance, the internal organs and the autonomic processes that regulate our bodies. At the other end are rational decisions—adaptive and species-altering across tens of thousands of generations in prehistoric epochs. It is at this end of the continuum, where rational choice and innate intuitions can overlap and reinforce one another, that we find important adaptations relevant to understanding the human personality, including the innate value systems implicit in morality, sociality, politics, religion, and the arts. Prehistoric choices honed the human virtues as we now know them: the admiration of altruism, skill, strength, intelligence, industriousness, courage, imagination, eloquence, diligence, kindness, and so forth.

The revelations of Darwin's later work—beautifully expli-
cated as well in books by Helena Cronin, Amotz and Avishag
Zahavi, and Geoffrey Miller—have completely altered my thinking about the development of culture. It is not just survival in
a natural environment that has made human beings what they
are. In terms of our personalities, we are, strange to say, a self-
made species. For me this is a genuine revelation, as it puts in a
new genetic light many human values that have hitherto been
regarded as purely cultural.

LINDA S. GOTTFREDSON

Sociologist, University of Delaware; codirector of the Project for the Study of Intelligence and Society

Human Innovation Fueled Human Evolution

For an empiricist, science brings many surprises. It has continued to change my thinking about many phenomena by challenging my presumptions about them. Among the first of my assumptions to be felled by evidence was that career choice proceeds in adolescence by identifying one's most preferred options; it actually begins early in childhood as a taken-for-granted process of eliminating the least acceptable from further consideration. Another mistaken presumption was that different abilities would be important for performing well in different occupations. The notion that any single ability (e.g., IQ) could predict performance to an appreciable degree in all jobs seemed far-fetched the first time I heard it, but that's just what my own attempt to catalog the predictors of job performance would help confirm. My root error had been to assume that different cognitive abilities (verbal, quantitative, etc.) are independent, that, in today's terms, there are "multiple intelligences." Empirical evidence says otherwise.

The most difficult ideas to change are those which seem so obviously true that we can scarcely imagine otherwise until confronted with unambiguous disconfirmation. For example, even behavioral geneticists had long presumed that nongenetic influences on intelligence and other human traits grow with age, while genetic ones weaken. Evidence reveals the opposite for intelligence and perhaps other human traits as well: Heritabilities actually increase with age. My attempt to explain the evolution of high human intelligence has also led me to question another such "obvious truth"—namely, that human evolution ceased when man took control of his environment. I now suspect that precisely the opposite occurred. Here is why.

Human innovation itself may explain the rapid increase in human intelligence during the last five hundred thousand years. Although it has improved the average lot of humankind, innovation creates evolutionarily novel hazards that put the less intelligent members of a group at relatively greater risk of accidental injury and death. Consider the first and perhaps most important human innovation, the controlled use of fire. It is still a major cause of death worldwide, as are falls from man-made structures and injuries from tools, weapons, vehicles, and domesticated animals. Much of humankind has indeed escaped from its environment of evolutionary adaptation, but only by fabricating new and increasingly complicated physical ecologies. Brighter individuals are better able not only to extract the benefits of successive innovations but also to avoid the novel threats to life and limb they create. Unintentional injuries and deaths have such a large chance component and their causes are so varied that we tend to dismiss them as mere accidents—as if they were uncontrollable. Yet all are to some extent preventable with foresight or effective response, which gives an edge to the more intelligent individuals. Evolution requires only tiny such differences in odds of survival in order to ratchet up intelligence over thousands of generations. If human innovation fueled human evolution in the past, then it likely still does today.

Another of my presumptions bit the dust, but in the process exposed a more fundamental, long-brewing challenge to my thinking about scientific explanation. At least in the social sciences, we seek big effects when predicting human behavior, whether we are trying to explain differences in happiness, job performance, depression, health, or income. "Effect size" (percentage of variance explained, standardized mean difference, and so on) has become our yardstick for judging the substantive importance of potential causes. Yet, while strong correlations between individuals' attributes and their fates may signal causal importance, small correlations do not necessarily signal unimportance.

Evolution provides an obvious example. Like the house in a gambling casino, evolution realizes big gains by playing small odds over myriad players and long stretches of time. The small-is-inconsequential presumption is so ingrained and reflexive, however, that even those of us who seek to explain the evolution of human intelligence over the eons have often rejected hypothesized mechanisms (say, superior hunting skills) when they could not explain differential survival or reproductive success within a single generation.

IQ tests provide a useful analogy for understanding the power of small but consistent effects. No single IQ test item measures intelligence well or has much predictive power. Yet, with enough items, one gets an excellent test of general intelligence (a.k.a. g) from only weakly g-loaded items. How? When test items are considered one by one, the role of chance dominates in determining who answers the item correctly. When test takers' responses to many such items are added together, however, the random effects tend to cancel one another out, and g's small contribution to all answers piles up. The result is a test that measures almost nothing but g.

I have come to suspect that some of the most important forces shaping human populations work in this inconspicuous but inexorable manner. When operating in individual instances, their impact is so small as to seem inconsequential, yet their consistent impact over events or individuals produces marked effects. To take a specific example, only the calculus of small but consistent tendencies in health behavior over a lifetime, not just accidental death, seems likely to explain many demographic disparities in morbidity and mortality.

Developing techniques to identify, trace, and quantify such influences will be a challenge. It currently bedevils behavioral geneticists who, having failed to find any genes with substantial influence on intelligence (within the normal range of variation), are now formulating strategies to identify genes that may account for at most only 0.5 percent of the variance in intelligence.

Social and technological network topology researcher; adjunct professor, Interactive Telecommunications Program, New York University; author of Here Comes Everybody: The Power of Organizing Without Organizations

The Doctrine of Joint Belief

I was a science geek with a religious upbringing—an Episcopalian upbringing, to be precise, which is pretty weak tea as far as pious fervor goes. Raised in this tradition, I learned, without ever being explicitly taught, that religion and science were compatible. My people had no truck with "young Earth" creationism or antievolutionary cant, thank you very much, and if some people's views clashed with scientific discovery, well, that was their fault for being so fundamentalist.

Since we couldn't rely on the literal truth of the Bible, we needed a fallback position to guide our views on religion and science. That position was what I'll call the Doctrine of Joint Belief: "Noted Scientist X has accepted Jesus as Lord and Savior. Therefore, religion and science are compatible." (Substitute deity to taste.) You can still see this argument today, where the beliefs of Francis Collins or Freeman Dyson, both accomplished scientists, are held up as evidence of such compatibility.

Belief in compatibility is different from belief in God. Even after I stopped believing, I thought religious dogma, though incorrect, was not directly incompatible with science (a view sketched out by Stephen Gould as "nonoverlapping magisteria"). I've now changed my mind, for the obvious reason: I was wrong. The idea that religious scientists demonstrate that religion and science are compatible is ridiculous, and I'm embarrassed that I ever believed it. Having believed for so long, however, I understand its attraction, and its fatal weaknesses.

The Doctrine of Joint Belief isn't evidence of harmony between two systems of thought. It simply offers permission to ignore the clash between them. Skeptics aren't convinced by the doctrine, unsurprisingly, because it offers no testable proposition. What is surprising is that its supposed adherents don't believe it either. If joint beliefs were compatible beliefs, there could be no such thing as heresy. Christianity would be compatible not just with science but with astrology (roughly as many Americans believe in astrology as believe in evolution), racism (because of the churches who use the "curse of Ham" to justify racial segregation), and on through the list of every pair of beliefs held by practicing Christians.

To get around this, one could declare that, for some arbitrary reason, the coexistence of beliefs is relevant only to questions of religion and science but not to astrology or anything else. Such a stricture doesn't strengthen the argument, however, because an appeal to the particular religious beliefs of scientists means having to explain why the majority of them are atheists. (See the 1998 Larson and Witham study for the numbers.)* Picking out the minority who aren't atheists and holding only them up as exemplars is simply special pleading (not to mention lousy statistics).

---

* Edward J. Larson and Larry Witham, "Leading Scientists Still Reject God," *Nature*, July 23, 1998; Vol. 394, p. 313.

The works that changed my mind about compatibility were Pascal Boyer's *Religion Explained* and Scott Atran's *In Gods We Trust*; both lay out the ways in which religious belief is a special kind of thought, incompatible with the kind of skepticism that makes science work. In Boyer and Atran's view, religious thought doesn't simply happen to be false; being false is the point, the thing that makes belief both memorable and effective. Psychologically, we overcommit to the ascription of agency, even when dealing with random events (confirmation can be had in any casino). Belief in God rides in on that mental eagerness, in the same way optical illusions ride in on our tendency to overinterpret ambiguous visual cues. Sociologically, the adherence to what Atran diplomatically calls "counterfactual beliefs" serves both to create and advertise in-group commitment among adherents. Anybody can believe in things that are true, but it takes a lot of coordinated effort to get people to believe in virgin birth or resurrection of the dead.

We are early in one of the periodic paroxysms of conflict between faith and evidence. I suspect this conflict will restructure society, as happened after Galileo, rather than leading to a quick truce, as after Scopes—not least because the global tribe of atheists now have a medium in which they can discover one another and refine and communicate their message.

One of the key battles is to insist on the incompatibility of beliefs based on evidence and beliefs that ignore evidence. Saying that the mental lives of a Francis Collins or a Freeman Dyson prove that religion and science are compatible is like saying that the sex lives of Bill Clinton or Ted Haggard prove that marriage and adultery are compatible. The people we need to watch out for in this part of the debate aren't the fundamentalists, they're the moderates, the ones who think that if religious belief is made metaphorical enough, incompatibility with science can be waved away. It can't be, and we need to say so, especially to the people like me, before I changed my mind.

RANDOLPH M. NESSE

Psychiatrist, University of Michigan; coauthor (with George C. Williams) of Why We Get Sick: The New Science of Darwinian Medicine

Universities and the Pursuit of Truth

I used to believe that you could find out what is true by finding the smartest people and finding out what they think. However, the most brilliant people keep turning out to be wrong. Linus Pauling's ideas about Vitamin C are fresh in mind, but the famous physicist Lord Kelvin did more harm in 1900, with his calculations based on the rate of Earth's cooling that seemed to show that there had not been enough time for evolution to take place. A lot of the belief that smart people are right is an illusion caused by smart people being very convincing—even when they are wrong.

I also used to believe that you could find out what is true by relying on experts—smart experts—who devote themselves to a topic. But most of us remember being told to eat margarine because it is safer than butter; then it turned out that trans fats are worse. Doctors told women to use hormone replacement therapy (HRT) to prevent heart attacks, but HRT turned out to increase heart attacks. Even when they're not wrong, expert reports often don't tell you what is true. For instance, read reviews by experts about antidepressants; they provide reams of data, but you won't often find the simple conclusion that these drugs are not all that helpful for most patients. It is not just others; I shudder to think about all the false beliefs I have unknowingly but confidently passed on to my patients, thanks to my trust in experts. Everyone should read the *PLoS Medicine* article by John Ioannidis: "Why Most Published Research Findings Are False."

Finally, I used to believe that truth had a special home in universities. After all, universities are supposed to be devoted to finding out what is true and teaching students what we know and how to find things out for themselves. Universities may be the best show in town for truth pursuers, but most of them stifle innovation and constructive engagement of real controversies — not just sometimes but most of the time, systematically.

How can this be? Everyone is trying so hard to encourage innovation! The regents take great pains to find a president who supports integrity and creativity; the president chooses exemplary deans, who mount massive searches for the best department chairs; those chairs often hire supporters who work in their own areas — but what if one wants to hire someone doing truly innovative work, someone who might challenge established opinions? Faculty committees intervene to ensure that most positions go to people just about like themselves, and the dean asks how much grant funding for overhead a new faculty member will bring in. No one with new ideas, much less working in a new area or critical of established dogmas, can hope to get through this fine sieve. If they do, review committees are waiting. And so, by a process of unintentional selection, diversity of thought and topic is excluded. If it sneaks in, it is purged. The disciplines become ever more insular, and universities find themselves unwittingly inhibiting progress and genuine intellectual engagement. University leaders recognize this and hate it, so they are constantly creating new initiatives to foster innovative interdisciplinary work. These have the same lovely sincerity as new diets for the New Year and the same blindness to the structural factors responsible for the problems.

Where can we look to find what is true? Smart experts in universities are a place to start, but if we could acknowledge how hard it is for truth and its pursuers to find safe university lodgings, and how hard it is for even the smartest experts to offer objective conclusions, we could begin to design new social structures that would support real intellectual innovation and engagement.

DAVID GELERNTER

Computer scientist, Yale University; national fellow, American Enterprise Institute; author of Americanism: The Fourth Great Western Religion

Big Ideas Need Time to Soak In

What I've changed my mind about is the U.S. public's attitude to technology. I now understand that the public is cautious but not reactionary. I'll give two examples from my own experience. Both constitute long-term ideas of mine and so this piece might seem like an exercise in self-promotion, but my point is this: New ideas can take root and thrive, if we are willing to be patient and we refuse to be taken in by the myth of the "fast-moving world of technology."

I first described a GUI called Lifestreams in the *Washington Post* in 1994. By the early 2000s, I thought this system was dead in the water, destined to be resurrected in a grad student's footnote around the twenty-ninth century. The problem was (I thought) that Lifestreams was too unfamiliar, insufficiently "evolutionary" and too "revolutionary" (as the good folks at ARPA like to say). Unless you go step-by-step with the public and the industry, you lose.

But today "lifestreams" are all over the Net (take a look yourself), and I'm told that "lifestreaming" has turned into a verb at some recent Internet conferences. According to ZDnet. com: "Basically what's important about the OLPC [one laptop per child] has nothing to do with its nominal purposes and everything to do with its interface. Ultimately traceable to David Gelernter's Lifestreams model, this is not just a remake of Apple's evolution of the original work at Palo Alto but something new."

Moral: The public may be cautious, but it is not reactionary.

In a 1991 book called *Mirror Worlds*, I predicted that everyone would be putting their personal stuff in the cybersphere (a.k.a. "the clouds"). I said the same in a 2000 manifesto on *Edge* called "The Second Coming," and in various other pieces in between. By 2005 or so, I assumed that once again I'd jumped the gun, by too long to learn the results pre-posthumously—but once again this (of all topics) turns out to be hot and all over the place nowadays. "Cloud computing" is the next big thing. What does this all prove? If you're patient, good ideas find audiences. But you have to be very patient.

And if you expect to cash in on long-term ideas in the United States, you're certifiable.

This last point is a lesson I teach my students, and on this item I haven't changed (and don't expect to change) my mind. Universities used to believe that good and useful ideas were their own reward, whether you made any money or not. That's still a good worldview.

Science editor of The Times *(London)*

Consultations Properly Run

I used to take the view that public consultations about science policy were pointless. While the idea of asking ordinary people's opinions about controversial research sounds reasonable, it is astonishingly difficult to do well.

When governments canvass about issues such as biotechnology or embryo research, what usually happens is that the whole exercise gets captured by special interests. A vocal minority with strong opinions that are already widely known and impervious to argument — think Greenpeace and the embryo-rights lobby — get their responses in early and often. The much larger proportion of people who consider themselves neutral, open to persuasion, uninformed, or uninterested rarely bother to take part. Public opinion is then deemed to have spoken, without reflecting true public opinion at all. Wouldn't it be better, I thought, to let scientists get on with their research, subject to occasional oversight by specialist panels with appropriate ethical expertise?

Well, to a point. Public consultations can indeed be worse than useless, particularly when the British Government has done the consulting; its exercises on genetically modified crops and embryo-research laws were particularly ill-judged. As Sir David King said recently, they have taught us what not to do. Their failure, though, has stimulated some interesting thinking that has convinced me that it is possible to engage ordinary people in quite complex scientific issues, without letting the usual suspects shout everybody else down.

The Human Fertilisation and Embryology Authority's recent work on cytoplasmic hybrid embryos is a case in point. The traditional part of the exercise had familiar results: Pro-lifers and anti–genetic engineering groups mobilized, so 494 of the 810 written submissions were hostile. Careful questioning, however, established that almost all these came from people who oppose all embryo research in all circumstances. A more scientific poll found 61 percent backing for interspecies embryos, if these were to be used for medical research. Detailed deliberative workshops revealed that once the rationale for the experiments was properly explained, large majorities overcame "instinctive repulsion" and supported the work.

If consultations are properly run in this way, there is a lot to be said for them. They can actually build public understanding of potentially controversial research, and shoot the fox of science's shrillest critics.

In many ways, they are rather more helpful than seeking advice from bioethicists, whose importance to ethical research I've increasingly come to doubt. It's not that philosophy of science is not a worthwhile academic discipline—it can be stimulating and thought-provoking. The problem is that a bioethicist can almost always be found to support any position. Leon Kass and John Harris are both eminent bioethicists, yet the counsel you would expect them to give on embryo-research laws is going to be rather different.

Politicians—or scientists—can and do deliberately appoint ethicists according to their preexisting worldviews, then trumpet their advice as somehow independent and authoritative, as if their subject were physics. If specialist bioethics has a role to play in regulation of science, it is in framing the questions that researchers and the public at large should consider. It can't just be a fig leaf for decisions people were always going to make anyway.

Founder and CEO of O'Reilly Media, Inc.

The Social Graph as the Next Big Thing

In November 2002, Clay Shirky organized a "social software summit," based on the premise that we were entering a "golden age of social software . . . greatly extending the ability of groups to self-organize."

I was skeptical of the term "social software" at the time. The explicit social software of the day—applications like Friendster and Meetup—were interesting, but didn't seem likely to be the seed of the next big Silicon Valley revolution.

I preferred to focus instead on the related ideas that I eventually formulated as Web 2.0—namely, that the Internet is displacing Microsoft Windows as the dominant software development platform and that the competitive edge on that platform comes from aggregating the collective intelligence of everyone who uses the platform. The common thread that linked Google's PageRank, eBay's marketplace, Amazon's user reviews, Wikipedia's user-generated encyclopedia, and craigslist's self-service classified advertising seemed too broad a phenomenon to be successfully captured by the term "social software." (This is also my complaint about the term "user-generated content.") By framing the phenomenon too narrowly, you can exclude the exemplars that help us to understand its true nature. I was looking for a bigger metaphor, one that would tie together everything from open-source software to the rise of Web applications.

You wouldn't think to describe Google as social software, yet Google's search results are profoundly shaped by its collective interactions with its users: Every time someone makes a link on the Web, Google follows that link to find the new site. It weights the value of the link based on a kind of implicit social graph (a link from site A is more authoritative than one from site B, based in part on the size and quality of the network that in turn references either A or B). When someone makes a search, they also benefit from the data Google has mined from the choices millions of other people have made when following links provided as the result of previous searches.

You wouldn't describe eBay or craigslist or Wikipedia as social software either, yet each of them is the product of a passionate community, without which none of those sites would exist, and from which they draw their strength, like Antaeus touching Mother Earth. Photo-sharing site Flickr or bookmark-sharing site del.icio.us (both now owned by Yahoo!) also exploit the power of an Internet community to build a collective work that is more valuable than could be provided by an individual contributor. But again, the social aspect is implicit—harnessed and applied, but never the featured act.

Now, five years after Clay's social software summit, Facebook, an application that explicitly explores the notion of the social network, has captured the imagination of those looking for the next Internet frontier. I find myself ruefully remembering my skeptical comments to Clay after the summit and wondering if he's saying, "I told you so."

Mark Zuckerberg, Facebook's young founder and CEO, woke up the industry when he began speaking of "the social graph"—that's computer-science-speak for the mathematical structure that maps the relationships between people participating in Facebook—as the core of his platform. There is real power in thinking of today's leading Internet applications explicitly as social software.

Mark's insight that the opportunity is not just about building a social-networking site but rather building a platform based on the social graph itself provides a lens through which to rethink countless other applications. Products like Xobni (inbox spelled backward) and MarkLogic's MarkMail explore the social graph hidden in our e-mail communications; Google and Yahoo! have both announced projects around this same idea. Google also acquired Jaiku, a pioneer in building a social-graph-enabled address book for the phone.

This is not to say that the idea of the social graph as the next big thing invalidates the other insights I was working with. Instead, it clarifies and expands them:

- Massive collections of data and the software that manipulates those collections, not software alone, are the heart of the next generation of applications.
- The social graph is only one instance of a class of data structure that will prove increasingly important as we build applications powered by data at Internet scale. You can think of the mapping of people, businesses, and events to places as the "location graph," or the relationship of search queries to results and advertisements as the "question-answer graph."
- The graph exists outside of any particular application; multiple applications may explore and expose parts of it, gradually building a model of relationships that exist in the real world.

- As these various data graphs become the indispensable foundation of the next generation "Internet operating system," we face one of two outcomes: Either the data will be shared by interoperable applications or the company that first gets to a critical mass of useful data will become the supplier to other applications and, ultimately, the master of that domain.

So have I really changed my mind? As you can see, I'm incorporating "social software" into my own ongoing explanations of the future of computer applications.

It's curious to look back at the notes from that first social software summit. Many core insights are there, but the details are all wrong. Many of the projects and companies mentioned have disappeared, while the ideas have moved beyond that small group of thirty or so people and, in the process, have become clearer and more focused, imperceptibly shifting from what we thought then to what we think now.

Both Clay, who thought then that "social software" was a meaningful metaphor, and I, who found it less useful then than I do today, have changed our minds. A concept is a frame, an organizing principle, a tool that helps us see. It seems to me that we all change our minds every day through the accretion of new facts, new ideas, new circumstances. We constantly retell the story of the past as seen through the lens of the present, and only sometimes are the changes profound enough to require a complete repudiation of what went before.

Ideas themselves are perhaps the ultimate social software, evolving via the conversations we have with each other, the artifacts we create, and the stories we tell to explain them.

Yes, if facts change our mind, that's science. But when ideas change our minds, we see those facts afresh, and that's history, culture, science, and philosophy all in one.

DAVID GOODHART

Founder and editor of Prospect *magazine*

The Specialness of Citizenship

The nation-state is too big for the local things, too small for the international things, and the root of most of the world's ills. This was a central part of liberal baby-boomer common sense when I was growing up, especially if you came from a (still) dominant country like Britain. Moreover, to show any sense of national feeling—apart from contempt for your national traditions—was a sign that you lacked political sophistication.

I now believe this is mainly nonsense. Nationalism can, of course, be a destructive force, and we were growing up in the shadow of its nineteenth- and twentieth-century excesses. In reaction to that, most of the civilized world had, by the mid-twentieth century, signed on to a liberal universalism (as embodied in the 1948 U.N. Universal Declaration of Human Rights) that stressed the moral equality of all humans. I am happy to sign on to that, too, of course, but I no longer see that commitment as necessarily conflicting with belief in the nation-state. Indeed I think many antinational liberals make a sort of category error; belief in the moral equality of all humans does not mean that we have the same obligations to all humans. Membership in the political community of a modern nation-state places onerous duties on us, to obey laws and pay taxes, but also grants us many rights and freedoms—and they make our fellow citizens politically special to us in a way that citizens of other countries are not. This fellow citizen favoritism is most vividly illustrated in the factoid that every year in Britain we spend twenty-five times more on the National Health Service than we do on development aid.

Moreover, if the nation-state can be a destructive force, it is also at the root of what many liberals hold dear: representative democracy, accountability, the welfare state, redistribution of wealth, and the very idea of equal citizenship for people of all colors and creeds. Few of these things have worked to any significant extent beyond the confines of the nation-state, which is not to say that they couldn't at some point in the future. (Indeed, they already do so, to a small extent, in the European Union.) If you look around at the daily news—contested elections in Kenya, death in Pakistan—most of the bad news these days comes from too little nation-state, not too much. And why was rapid economic development possible in the Asian tigers but not in Africa? Surely the existence of well-functioning nation-states and a strong sense of national solidarity in the tigers had something to do with it.

And in rich Western countries, as other forms of human solidarity—social class, religion, ethnicity, and so on—have been replaced by individualism and narrower group identities, holding on to some sense of national solidarity remains more important than ever to the good society. Empathy toward strangers who are fellow citizens (and with whom one shares history, institutions, and social and political obligations) underpins successful modern states, but this need not stand in the way of empathy toward all humans. It just remains true that charity begins at home.

Scholar of randomness and former trader/applied statistician; author of The Black Swan: The Impact of the Highly Improbable

The Fallacy of Probability

I spent a long time believing in the centrality of probability in life and advocating that we should express everything in terms of degrees of credence, with unitary probabilities as a special case for total certainties and null for total implausibility. Critical thinking, knowledge, beliefs—everything needed to be probabilized. Until I came to realize, twelve years ago, that I was wrong in this notion that the calculus of probability could be a guide to life and help society. Indeed, it is only in very rare circumstances that probability (by itself) is a guide to decision making. It is a clumsy academic construction, extremely artificial, and nonobservable. Probability is backed out of decisions; it is not a construct to be handled in a stand-alone way in real-life decision making. It has caused harm in many fields.

Consider the following statement. "I think that this book is going to be a flop, but I would be very happy to publish it." Is the statement incoherent? Of course not: Even if the book is very likely to be a flop, it may make economic sense to publish it (for someone with deep pockets and the right appetite), since one cannot ignore the small possibility of a handsome windfall or the even smaller possibility of a huge windfall. We can easily see that when it comes to low odds, decision making no longer depends on the probability alone. It is the pair—probability times payoff (or a series of payoffs)—the expectation, that matters. On occasion, the potential payoff can be so vast that it dwarfs the probability—and these are usually real-world situations in which probability is not computable.

Consequently, there is a difference between knowledge and action. You cannot naively rely on scientific statistical knowledge (as they define it) or what the epistemologists call justified true belief for nontextbook decisions. Statistically oriented modern science is typically based on Right/Wrong, with a set confidence level, stripped of consequences. Would you take a headache pill if it was deemed effective at a 95 percent confidence level? Most certainly. But would you take the pill if it is established that it is "not lethal" at a 95 percent confidence level? I hope not.

When I discuss the impact of the highly improbable ("black swans"), people make the automatic mistake of thinking that the message is that these "black swans" are necessarily more probable than assumed by conventional methods. They are mostly less probable. Consider that, in a winner-take-all environment, such as the arts, the odds of success are low, since there are fewer successful people, but the payoff is disproportionately high. So, in a fat-tailed environment (what I call Extremistan), rare events are less frequent (their probability is lower), but they are so effective that their contribution to the total pie is more substantial.

[Technical note: The distinction is, simply, between raw probability, $P[x>K]$, i.e., the probability of exceeding K, and $E[x|x>K]$, the expectation of x conditional on $x>K$. It is the difference between the zero[th] moment and the first moment. The latter is what usually matters for decisions. And it is the (conditional) first moment that needs to be the core of decision making. What I saw in 1995 was that an out-of-the-money option value increases when the probability of the event decreases, making me feel that everything I thought until then was wrong.]

What causes severe mistakes is that outside the special cases of casinos and lotteries, you almost never face a single probability with a single (and known) payoff. You may face, say, a 5 percent probability of an earthquake of magnitude three or higher, a 2 percent probability of one of magnitude four or higher, and so forth. The same with wars: You have a risk of different levels of damage, each with a different probability. "What is the probability of war?" is a meaningless question for risk assessment.

So it is wrong to look just at a single probability of a single event in cases of richer possibilities (like focusing on such questions as "What is the probability of losing a million dollars?" while ignoring that, conditional on losing more than a million dollars, you may have an expected loss of twenty million dollars, a hundred million dollars, or just one million dollars). Once again, real life is not a casino with simple bets. This is the error that helps the banking system go bust with an astonishing regularity. I've shown that institutions that are exposed to negative black swans—such as banks and some classes of insurance ventures—have almost never been profitable over long periods. The problem of the illustrative current subprime mortgage mess is not so much that the "quants" and other pseudo-experts in bank risk-management were wrong about the probabilities (they were) but that they were severely wrong about the different layers of depth of potential negative outcomes. For instance, Morgan Stanley has lost about ten billion dollars (so far), while allegedly having foreseen a subprime crisis and executed hedges against it; they just did not realize how deep it would go and had open exposure to the big tail risks. This is routine. A friend who went bust during the crash of 1987 told me, "I was betting that it would happen, but I did not know it would go that far."

The point is mathematically simple, but does not register easily. I've enjoyed giving math students the following quiz (to be answered intuitively, on the spot). In a Gaussian world, the probability of exceeding one standard deviation is around 16 percent. What are the odds of exceeding it under a distribution of fatter tails (with same mean and variance)? The right answer: lower, not higher—the number of deviations drops, but the few that take place matter more. It was entertaining to see that most of the graduate students got it wrong. Those who are untrained in the calculus of probability have a far better intuition of these matters.

Another complication is that just as probability and payoff are inseparable, so one cannot extract another complicated component—utility—from the decision-making equation. Fortunately the ancients, with all their tricks and accumulated wisdom in decision making, knew a lot of that—at least, better than modern-day probability theorists. Let us stop systematically treating them as if they were idiots. Most texts blame the ancients for their ignorance of the calculus of probability: The Babylonians, Egyptians, and Romans in spite of their engineering sophistication, and the Arabs in spite of their taste for mathematics, were blamed for not having produced a calculus of probability (the latter being, incidentally, a myth, since Umayyad scholars used relative word frequencies to determine authorships of holy texts and decrypt messages). The reason was foolishly attributed to theology, lack of sophistication, lack of something people call the "scientific method," or belief in fate. The ancients just made decisions in a more ecologically sophisticated manner than modern epistemology-minded people. They integrated skeptical Pyrrhonian empiricism into decision making. As I said, consider that belief (i.e., epistemology) and action (i.e., decision making), the way they are practiced, are largely not consistent with each other.

Let us apply the point to the current debate on carbon emissions and climate change. Correspondents keep asking me if the climate worriers are basing their claims on shoddy science and whether, owing to nonlinearities, their forecasts are marred with such resultant errors we should ignore them. Now, even if I agreed that it was shoddy science, even if I agreed with the statement that the climate folks were most probably wrong, I would still opt for the most ecologically conservative stance: Leave Planet Earth the way we found it. Consider the consequences of the very remote possibility that they may be right—or, worse, the even more remote possibility that they may be extremely right.

BART KOSKO

Information scientist, University of Southern California; author of Noise

The Sample Mean

I have changed my mind about using the sample mean as the best way to combine measurements into a single predictive value. Sometimes it's the best way to combine data, but in general you don't know that in advance. So it is not the one number from or about a data set that I would want to know in the face of total uncertainty if my life depended on the predicted outcome.

Using the sample mean always seemed like the natural thing to do: Just add up the numerical data and divide by the number of data. I don't recall ever doubting that procedure until my college years. Even then, I kept running into the mean in science classes and even in philosophy classes, where the discussion of ethics sometimes revolved around Aristotle's theory of the golden mean. There were occasional mentions of medians and modes and other measures of central tendency, but they were only occasional.

The sample mean also kept emerging as the optimal way to combine data in many formal settings. At least, it did given what appeared to be the reasonable criterion of minimizing the squared errors of the observations. The sample mean falls out from just one quick application of the differential calculus. So the sample mean not only had on its side mathematical proof and the resulting prominence of appearing in hundreds, if not thousands, of textbooks and journal articles but it also was, and remains, the evidentiary workhorse of modern applied science and engineering. The sample mean summarizes test scores and gets plotted in trend lines and centers confidence intervals among numerous other applications.

Then I ran into the counterexample of Cauchy data. These data come from bell curves with tails just slightly thicker than the familiar "normal" bell curve. Cauchy bell curves also describe "normal" events that correspond to the main bell of the curves. But Cauchy bell curves have thicker tails than normal bell curves have, and these thicker tails allow for many more "outliers," or rare events. And Cauchy bell curves arise in a variety of real and theoretical cases. The counterexample is that the sample mean of Cauchy data does not improve no matter how many samples you combine. This result contrasts with the usual result from sampling theory—that the variance of the sample mean falls with each new measurement and hence predictive accuracy improves with sample size (assuming that the square-based variance term measures dispersion and that such a mathematical construct always produces a finite value—which it need not produce in general). The sample mean of ten thousand Cauchy data points has no more predictive power than does the sample mean of ten such data points. Indeed, the sample mean of Cauchy data has no more predictive power than does any one of the data points picked at random. This counterexample is but one of the anomalous effects that arise from averaging data from many real-world probability curves that deviate from the normal bell curve or from the twenty or so other closed-form probability curves that have found their way into the literature in the last century.

Nor have scientists always used the sample mean. Historians of mathematics have pointed to the late sixteenth century and the introduction of the decimal system for the start of the modern practice of computing the sample mean of data sets to estimate typical parameters. Before then, the mean apparently meant the arithmetic average of just two numbers, as it did with Aristotle. So Hernán Cortés may well have had a clear idea about the typical height of an adult male Aztec in the early sixteenth century. But he quite likely did not arrive at his estimate of the typical height by adding measured heights of Aztec males and then dividing by the number added. We have no reason to believe that Cortés would have resorted to such a computation if the Church or King Charles had pressed him to justify his estimate. He might just as well have lined up a large number of Aztec adult males from shortest to tallest and then reported the height of the one in the middle.

There was a related and deeper problem with the sample mean: It is not robust. Extremely small or large values distort it. This rotten-apple property stems from working not with measurement errors but with squared errors. The squaring operation exaggerates extreme data even though it greatly simplifies the calculus when trying to find the estimate that minimizes the observed errors. That estimate turns out to be the sample mean but not in general if one works with the raw error itself or other measures. The statistical surprise of sorts is that using the raw or absolute error of the data gives the sample median as the optimal estimate.

The sample median is robust against outliers. If you throw away the largest and smallest values in a data set, the median does not change, but the sample mean does (and gives a more robust, "trimmed" mean, as used in combining the judging scores in figure skating and elsewhere to remove judging bias). Realtors have long since stated typical housing prices as sample medians rather than sample means, because a few mansions can so easily skew the sample mean. The sample median would not change even if the price of the most expensive house rose to infinity. The median would still be the middle-ranked house if the number of houses were odd. But this robustness is not a free lunch. It comes at the cost of ignoring some of the information in the numerical magnitudes of the data and has its own complexities for multidimensional data.

Other evidence pointed to using the sample median rather than the sample mean. Statisticians have computed the so-called breakdown point of these and other statistical measures of central tendency. The breakdown point measures the largest proportion of data outliers that a statistic can endure before it breaks down in a formal sense of producing very large deviations. The sample median achieves the theoretical maximum breakdown point. The sample mean does not come close. The sample median also turns out to be the optimal estimate for certain types of data (such as Laplacian data) found in many problems of image processing and elsewhere—if the criterion is maximizing the probability or likelihood of the observed data. And the sample median can also center confidence intervals. So it, too, gives rises to hypothesis tests and does so while making fewer assumptions about the data than the sample mean often requires for the same task.

The clincher was the increasing use of adaptive or neural-type algorithms in engineering and especially in signal processing. These algorithms cancel echoes and noise on phone lines as well as steer antennae and dampen vibrations in control systems. The whole point of using an adaptive algorithm is that the engineer cannot reasonably foresee all the statistical patterns of noise and signals that will bombard the system over its lifetime. No type of lifetime average will give the kind of performance that real-time adaptation will give if the adaptive algorithm is sufficiently sensitive and responsive to its measured environment. The trouble is that most of the standard adaptive algorithms derive from the same old and nonrobust assumptions about minimizing squared errors, and thus they result in the use of sample means or related nonrobust quantities. So real-world gusts of data wind tend to destabilize them. That's a high price to pay just because in effect it makes nineteenth-century calculus computations easy and because such easy computations still hold sway in so much of the engineering curriculum. It is an unreasonably high price to pay in many cases where a comparable robust median-based system or its kin both avoids such destabilization and performs similarly in good data weather and does so for only a slightly higher computational cost. There is a growing trend toward using robust algorithms. But engineers still have launched thousands of these nonrobust adaptive systems into the stream of commerce in recent years. We do not know whether the social costs involved from using these nonrobust algorithms are negligible or substantial.

So if, under total uncertainty, I had to pick a predictive number from a set of measured data and if my life depended on it, I would now pick the median.

Physicist, computer scientist; chairman of Applied Minds, Inc.; author of The Pattern on the Stone

Do It Yourself

As a child, I was told that hot water freezes faster than cold water. This was easy to refute in principle, so I did not believe it. Many years later, I learned that Aristotle had described the effect in his *Meteorologica*:

> *The fact that the water has previously been warmed contributes to its freezing quickly: for so it cools sooner. Hence many people, when they want to cool hot water quickly, begin by putting it in the sun. So the inhabitants of Pontus when they encamp on the ice to fish (they cut a hole in the ice and then fish) pour warm water round their reeds that it may freeze the quicker, for they use the ice like lead to fix the reeds. (E. W. Webster translation)*

I was impressed as always by Aristotle's clarity, confidence, and specificity. Of course, I do not expect you to be convinced that it is true simply because Aristotle said so, especially since his explanation is that "warm and cold react upon one another by recoil." (Aristotle, like us, was very good at making up explanations to justify his beliefs.) Instead, I hope that you will have the pleasure of being convinced, as I was, by trying the experiment yourself.